RENEWABLE ENERGY: RESEARCH, DEVELOPMENT AND POLICIES

HYDROPOWER AND ENERGY POTENTIAL AT NON-POWERED DAMS

RENEWABLE ENERGY: RESEARCH, DEVELOPMENT AND POLICIES

Additional books in this series can be found on Nova's website under the Series tab.

Additional e-books in this series can be found on Nova's website under the e-books tab.

ENERGY SCIENCE, ENGINEERING AND TECHNOLOGY

Additional books in this series can be found on Nova's website under the Series tab.

Additional e-books in this series can be found on Nova's website under the e-books tab.

RENEWABLE ENERGY: RESEARCH, DEVELOPMENT AND POLICIES

HYDROPOWER AND ENERGY POTENTIAL AT NON-POWERED DAMS

KEVIN F. ANDERSON
AND
SANDRA A. COLLINS
EDITORS

New York

Library of Congress Cataloging-in-Publication Data

ISBN: 978-1-62257-711-8

Published by Nova Science Publishers, Inc. † New York

CONTENTS

PREFACE

The United States has produced clean, renewable electricity from hydropower for more than 100 years, but hydropower-producing facilities represent only a fraction of the infrastructure development that has taken place on the nation's waterways. In contrast to the roughly 2,500 dams, the United States has more than 80,000 non-powered dams (NPDs)—dams that do not produce electricity—providing a variety of services ranging from water supply to inland navigation. Importantly, many of the monetary costs and environmental impacts of dam construction have already been incurred at NPDs, so adding power to the existing dam structure can often be achieved at lower cost, with less risk, and in a shorter timeframe than development requiring new dam construction. This book discusses hydropower generation, government involvement at federal and nonfederal levels and the energy potential at non-powered dams.

Chapter 1 – Congress is examining numerous energy sources to determine their contribution to the nation's energy portfolio and the federal role in supporting these sources. Hydropower, the use of flowing water to produce electricity, is one such source. Conventional hydropower accounted for approximately 6% of total U.S. net electricity generation in 2010.

Hydropower has advantages and disadvantages as an energy source. Its advantages include its status as a continuous, or baseload, power source that releases minimal air pollutants during power generation relative to fossil fuels. Some of its disadvantages, depending on the type of hydropower plant, include high initial capital costs, ecosystem disruption, and reduced generation during low water years and seasons.

Hydropower project ownership can be categorized as federal or non-federal. The bulk of federal projects are owned and managed by the Bureau of

Reclamation and the U.S. Army Corps of Engineers. Nonfederal projects are licensed and overseen by the Federal Energy Regulatory Commission (FERC).

Considered by many to be an established energy source, hydropower is not always discussed alongside clean or renewable energy sources in the ongoing energy debate. However, hydropower proponents argue that hydropower is cleaner than some conventional energy sources, and point to recent findings that additional hydropower capacity could help the United States reach proposed energy, economic, and environmental goals. Others argue that the expansion of hydropower in the form of numerous small hydropower projects could have environmental impacts and regulatory concerns similar to those of existing large projects.

Congress faces several issues as it determines how hydropower fits into a changing energy and economic landscape. For example, existing large hydropower infrastructure is aging; many of the nation's hydropower generators and dams are over 30 years old. Proposed options to address this concern include increasing federal funding, utilizing alternative funding, privatizing federally owned dams, and encouraging additional small-capacity generators, among other options. Additionally, whether to significantly expand or encourage expansion of hydropower is likely to require congressional input due to the uncertainty surrounding the clean and renewable energy portfolio within power markets. Potential expansion of hydropower projects could take place by improving efficiency at existing projects or by building new projects, or both. Congressional support for this approach is evident in the House passage of the Bureau of Reclamation Small Conduit Hydropower Development and Rural Jobs Act of 2012 (H.R. 2842). Senate activity on this matter includes the Hydropower Improvement Act of 2011 (S. 629), which proposes to establish a grants program for increased hydropower production, and to amend the Federal Power Act (FPA) to authorize FERC to exempt electric power generation facilities on federal lands from the act's requirements, among other things. Another issue is the rate at which FERC issues licenses for nonfederal projects, which is slower than some find ideal. The licensing process can be delayed significantly as stakeholders and the approximately dozen federal and state agencies involved give their input. FERC responded by developing a more streamlined licensing process in 2003. Still, some object to "mandatory conditions" that federal agencies can place on new or renewed hydropower facilities. The 112th Congress has introduced roughly 25 bills regarding hydropower, a quarter of which are state- or site-specific legislation.

Chapter 2 - The United States has produced clean, renewable electricity from hydropower for more than 100 years, but hydropower-producing facilities represent only a fraction of the infrastructure development that has taken place on the nation's waterways. In contrast to the roughly 2,500 dams that provide 78 gigawatts (GW) of conventional and 22 GW of pumped-storage hydropower, the United States has more than 80,000 non-powered dams (NPDs)—dams that do not produce electricity—providing a variety of services ranging from water supply to inland navigation. Importantly, many of the monetary costs and environmental impacts of dam construction have already been incurred at NPDs, so adding power to the existing dam structure can often be achieved at lower cost, with less risk, and in a shorter timeframe than development requiring new dam construction. The abundance, cost, and environmental favorability of NPDs, combined with the reliability and predictability of hydropower, make these dams a highly attractive source for expanding the nation's renewable energy supply.

To better characterize this unique national resource, the U.S. Department of Energy (DOE) Wind and Water Power Program has undertaken a national-scale analysis of U.S. dams to determine the ability of NPDs to provide hydroelectric power. DOE's Oak Ridge National Laboratory (ORNL), with input from DOE's Idaho National Laboratory, quantified the potential capacity and generation available from adding power production capability to U.S. NPDs. Of the more than 80,000 NPDs throughout the U.S., 54,391 dams were analyzed, with remaining dams eliminated from consideration due to erroneous geographic information, or erroneous flow or drainage area attributes that could not be resolved and corrected through independent investigation of maps and records. Anecdotal information suggests that these dams with missing or erroneous information are likely to be relatively small or have low potential to produce hydroelectric energy. Dams with a reported height of less than five feet were also excluded from analysis. A thorough quality control and review process ensured that the 54,391 remaining NPDs were analyzed and characterized as accurately as possible. Figure ES-1 demonstrates the spatial and capacity potential distribution of the nation's NPDs. Electric generating capacities included in the report were calculated using the assumption that all water passing a facility would be available for conversion into electrical energy and that hydraulic head at the facility would remain constant. The analysis did not consider the economic feasibility of developing each unpowered facility. The assessment provides preliminary information for stakeholders (such as developers, municipal planners, and policymakers), who can further evaluate the potential to increase hydropower production at NPD sites. Developers could use the information provided in this assessment to focus on more detailed analysis of sites that demonstrate a reasonable potential for being developed.

Adding power to U.S. NPDs has the potential to add up to 12 GW (12,000 megawatts or MW) of new renewable capacity—a potential equivalent to increasing the size of the existing conventional hydropower fleet by 15%. A majority of this potential is concentrated in just 100 NPDs, which could contribute approximately 8 GW of clean, reliable hydropower; the top 10 facilities alone could add up to 3 GW of new hydropower. Eighty-one of the 100 top NPDs are U.S. Army Corps of Engineers (USACE) facilities, many of which, including all of the top 10, are navigation locks on the Ohio River, Mississippi River, Alabama River, and Arkansas River, as well as their major tributaries. This study also shows that dams owned by the U.S. Bureau of Reclamation hold the potential to add approximately 260 MW of capacity; the Bureau has also engaged in an effort to conduct a more detailed evaluation of its own facilities.

In: Hydropower and Energy Potential ... ISBN 978-1-62257-711-8
Editors: K. F. Anderson, S. A. Collins

Chapter 1

HYDROPOWER: FEDERAL AND NONFEDERAL INVESTMENT*

Kelsi Bracmort, Charles V. Stern and Adam Vann

SUMMARY

Congress is examining numerous energy sources to determine their contribution to the nation's energy portfolio and the federal role in supporting these sources. Hydropower, the use of flowing water to produce electricity, is one such source. Conventional hydropower accounted for approximately 6% of total U.S. net electricity generation in 2010.

Hydropower has advantages and disadvantages as an energy source. Its advantages include its status as a continuous, or baseload, power source that releases minimal air pollutants during power generation relative to fossil fuels. Some of its disadvantages, depending on the type of hydropower plant, include high initial capital costs, ecosystem disruption, and reduced generation during low water years and seasons.

Hydropower project ownership can be categorized as federal or nonfederal. The bulk of federal projects are owned and managed by the Bureau of Reclamation and the U.S. Army Corps of Engineers. Nonfederal projects are licensed and overseen by the Federal Energy Regulatory Commission (FERC).

---

* This is an edited, reformatted and augmented version of a Congressional Research Service publication, CRS Report for Congress R42579, prepared for Members and Committees of Congress, from www.crs.gov, dated June 26, 2012.

Considered by many to be an established energy source, hydropower is not always discussed alongside clean or renewable energy sources in the ongoing energy debate. However, hydropower proponents argue that hydropower is cleaner than some conventional energy sources, and point to recent findings that additional hydropower capacity could help the United States reach proposed energy, economic, and environmental goals. Others argue that the expansion of hydropower in the form of numerous small hydropower projects could have environmental impacts and regulatory concerns similar to those of existing large projects.

Congress faces several issues as it determines how hydropower fits into a changing energy and economic landscape. For example, existing large hydropower infrastructure is aging; many of the nation's hydropower generators and dams are over 30 years old. Proposed options to address this concern include increasing federal funding, utilizing alternative funding, privatizing federally owned dams, and encouraging additional small-capacity generators, among other options. Additionally, whether to significantly expand or encourage expansion of hydropower is likely to require congressional input due to the uncertainty surrounding the clean and renewable energy portfolio within power markets. Potential expansion of hydropower projects could take place by improving efficiency at existing projects or by building new projects, or both. Congressional support for this approach is evident in the House passage of the Bureau of Reclamation Small Conduit Hydropower Development and Rural Jobs Act of 2012 (H.R. 2842). Senate activity on this matter includes the Hydropower Improvement Act of 2011 (S. 629), which proposes to establish a grants program for increased hydropower production, and to amend the Federal Power Act (FPA) to authorize FERC to exempt electric power generation facilities on federal lands from the act's requirements, among other things. Another issue is the rate at which FERC issues licenses for nonfederal projects, which is slower than some find ideal. The licensing process can be delayed significantly as stakeholders and the approximately dozen federal and state agencies involved give their input. FERC responded by developing a more streamlined licensing process in 2003. Still, some object to "mandatory conditions" that federal agencies can place on new or renewed hydropower facilities. The 112th Congress has introduced roughly 25 bills regarding hydropower, a quarter of which are state- or site-specific legislation.

INTRODUCTION

Conventional hydropower[1] accounted for nearly 6% (260,203 gigawatt hours) of total net U.S. electricity generation and 7.5% (78 gigawatts) of net summer generating capacity in 2010.[2] The United States has considerable hydropower potential beyond what is already developed.[3] Although there is some interest in increasing the current level of hydropower investment, such investment remains limited, in part because of federal and nonfederal financial constraints, high uncertainty in electricity generation policies and markets (which may dissuade capital investment), and the environmental operating requirements required for hydropower projects. Some hydropower proponents are pursuing policies to reduce what they view as impediments to hydropower development. Other stakeholders prefer that hydropower investments only proceed when they are protective of other interests, including other water users and the aquatic environment and its species.

At issue for Congress is whether, and if so how, to change federal support for hydropower and what priority to give hydropower vis-à-vis other energy investments, and social and environmental concerns. The 112th Congress has introduced over 25 bills dealing with various aspects of hydropower—a quarter of which are state- or site-specific legislation.[4] Some Members of Congress want to provide further support for small hydropower and nonfederal hydropower at federal sites. For example, the House passed the Bureau of Reclamation Small Conduit Hydropower Development and Rural Jobs Act of 2012 (H.R. 2842), which would amend the Reclamation Power Act of 1939 to authorize the Secretary of the Interior to contract for the development of small conduit hydropower (1.5 megawatts or less) at Reclamation facilities; require that power privilege leases be offered first to an irrigation district or water users association operating or receiving water from the applicable work; and exempt small conduit hydropower development from the National Environmental Policy Act of 1969 (NEPA) in select cases, among other things.[5] Also, the Hydropower Improvement Act of 2011 (S. 629) proposes to establish a Department of Energy grants program for increased hydropower production, and amends the Federal Power Act (FPA) to authorize the Federal Energy Regulatory Commission (FERC) to exempt electric power generation facilities on federal lands from the act's requirements among other things.

The variety of legislative proposals and stakeholder opinions stems partly from differing views of the benefits and costs of hydropower development. Many have pointed out advantages and drawbacks of conventional hydropower. Cited advantages include its renewable energy status,[6] zero to minimal greenhouse gas emissions during operation, and high operational efficiency. Supporters also note its ability to generally serve as a reliable and flexible domestic energy source. Generally speaking, hydropower generation can be dispatched on relatively short notice, and can supplement shortfalls in generation. However, others note that conventional hydropower has high initial capital costs, can be detrimental to surrounding ecosystems (e.g., fish and wildlife), may not be reliable during low water years and seasons, and may disrupt recreational or scenic values.[7]

The legislative environment for hydropower can be abstruse. Whether the hydropower project investor is a federal or nonfederal entity dictates which laws apply. Several federal government agencies own and operate large hydropower projects, while other agencies administer the process by which nonfederal hydropower projects are built, maintained, and operated. Large federal hydropower projects are managed primarily by the U.S. Department of the Interior's Bureau of Reclamation (hereinafter referred to as Reclamation) and the U.S. Army Corps of Engineers (hereinafter referred to as the Corps). The power from these projects is marketed by the Department of Energy's Power Marketing Administrations (PMAs). FERC regulates investigation, construction, and operations of nonfederal hydropower projects as well as overseeing dam safety for nonfederal projects.

This report explains how the federal government is involved directly in hydropower generation at federal facilities and in the regulation of nonfederal hydropower generation; the focus is on current roles and processes and common concerns and questions about changing those roles.

BACKGROUND

Most of U.S. hydropower capacity is from conventional hydropower.[8] Conventional hydropower plants take three general forms: storage (or impoundment), run-of-river (or diversion), and pumped storage.[9] A storage plant uses a dam to store enough water in a reservoir so that, when released, it flows through a penstock to a turbine, spinning it, which in turn activates a generator to produce electricity.[10] A run-of-river plant directs a portion of a river through a canal or penstock to generate electricity without the need for a reservoir. A pumped storage facility stores energy by pumping water from a lower reservoir to an upper reservoir during off-peak hours; and during periods of higher electricity demand, the water is released to generate electricity.

Conventional hydropower is a significant contributor to the national electric power portfolio.[11] Hydropower produced 6%-9% of total U.S. electric generation between 1998 and 2009, depending on water availability.[12] Hydropower accounted for roughly 6% and 9% of the U.S. generation in 2010 and 2011, respectively.[13] Hydropower was the largest contributor to renewable electric power generation in 2011, followed by wind, and wood and wood-derived fuels.[14] The top hydropower-producing states in 2011 were western states—Washington, Oregon, and California.[15]

There is an ongoing debate about whether hydropower should be characterized as "renewable." It has historically been characterized as renewable because it is a replenishable resource. More recently, others have asserted that it is not renewable because of its size (e.g., in the case of large-scale projects) and environmental impact, particularly on ecosystems, and in some cases induced evaporation (particularly at reservoirs at low elevations in dry climates) and greenhouse gas emissions during construction. For purposes of this report, hydropower is discussed as renewable because it does not originate from a fossil fuel (i.e., its source is not finite).

The precise number of hydropower projects in the United States is unknown.[16] Different databases yield different results based on selected criteria. *Figure 1* shows the location of both nonfederal (FERC-regulated) hydropower projects and federal hydropower projects. Additionally, 90 nonfederal hydropower projects are licensed to operate at Corps dams, and 28 are licensed to operate at Bureau of Reclamation sites.[17]

The public sector (federal, cooperatives, and municipalities) owns the majority of hydropower capacity, while the private sector (e.g., private utility, private non-utility, and industrial entities) owns the majority of individual hydropower projects. According to a 2006 Idaho National Lab (INL) study, the private sector owns two-thirds of U.S. hydropower plants. However, private plants account for just over 25% of production capacity.[18] Similarly, the public sector owns roughly one-third of the hydropower plants; these plants are responsible for roughly three-quarters of capacity. The federal government tends to own mostly large hydropower (> 30 MW) plants. The private sector and cooperatives tend to own mostly small hydro (1-30 MW) and low power (< 1 MW) plants. Small hydro and low power plants constitute approximately 11% and less than 1% of total hydropower production capacity, respectively. Large hydropower plants are responsible for approximately 89% of total hydropower capacity.[19]

FEDERAL HYDROPOWER

The federal government owns and operates approximately half of all U.S. hydroelectric generating capacity. This capacity is principally at large multi-purpose dams owned and operated by the Corps or Reclamation. Combined, the Corps and Reclamation operate almost all federal hydroelectric dam capacity (91% of federal capacity).[20] Other federal entities operating hydroelectric generation facilities include the Tennessee Valley Authority, the Bureau of Indian Affairs, and the International Boundary and Water Commission. Other entities also own facilities that generate hydroelectric power, but they are relatively small. While the electricity generated by these facilities is owned by these federal agencies, the Power Marketing Administrations (PMAs), which are part of the U.S. Department of Energy, are generally responsible for selling and distributing this power.

Federal hydropower capacity varies substantially by state and region. For instance, 90% of all federal capacity is found in 13 states, and the majority of this capacity is in the West.[21] Washington and California contain the greatest federal hydroelectric capacity, while Oregon, Arizona, Montana, and Idaho also have significant federal hydropower capacity. In the East, New York, Georgia, South Carolina, and North Carolina have the most federal capacity.

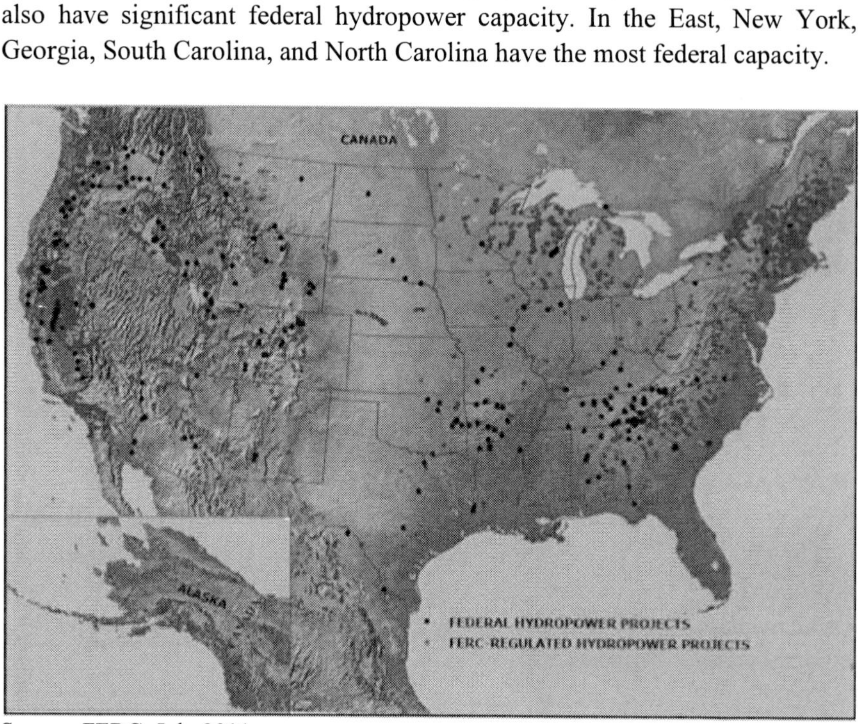

Source: FERC, July 2011.

Notes: According to FERC records, no licenses were issued for projects in Hawaii. However, one exemption was issued, and five projects in Hawaii have been issued preliminary permits.

Figure 1. Conventional U.S. Hydropower Projects: Federal and FERC-Regulated.

U.S. Army Corps of Engineers Projects

The Corps' multi-purpose dams are the largest producer of hydropower in the United States. Much of this power generation capacity is concentrated in the Pacific Northwest. The Corps constructed hydropower facilities at many of its water resources projects, beginning in 1925. Its most recent hydroelectric construction project was the R.D. Willis project in Texas, completed in 1989. Today, the Corps owns and operates 353 units at 75 projects, with a total estimated capacity of 25.8 GW, or approximately one-fourth of all national hydropower capacity.[22] These projects generate approximately 70,000 gigawatt hours (GWH) of hydropower annually, with an average gross revenue of $5 billion.[23] Revenues from the sale of this hydropower are deposited into the U.S. Treasury.

Hydropower generating units have a nominal 50-year life expectancy, and many Corps hydropower projects are nearing or exceeding this age. As of 2011, the median age of Corps hydropower facilities was 47 years, and 90% of Corps projects are 34 years or older.[24] Although hydroelectric generation is highly variable and depends on a number of factors, some have observed that nationwide, generation at Corps units has fallen over the last decade. According to the Corps, from 2000 to 2008 generation at its facilities fell from 71,600 GWH in 2000 to 61,700 GWH in 2008. The extent to which this trend is due to aging infrastructure in uncertain; however, some contend age is a factor in declining generation.[25] Relative to national generation totals, the Corps contribution to total hydropower generation decreased from approximately 27% of total production in 2000 to 24% of total production in 2008.[26]

In addition to overall declining generation trends, the delivery of Corps hydroelectric power is also an issue connected to aging infrastructure. That is, unit availability is down because of maintenance and repairs.[27] According to the Corps, many of its hydropower assets have fallen below the generally accepted hydropower industry goal of 95% unit availability. Unit availability at Corps dams has fallen to the point where no Corps division is meeting the 95% target, while at the same time, the total hours of "forced outages" have grown.[28]

Concerns related to aging infrastructure have resulted in internal reviews of the Corps hydropower program and recommendations for ways to improve Corps hydropower operations. The Corps has conducted preliminary reviews of foregone benefits associated with its aging hydropower infrastructure under its Hydropower Modernization Initiative (HMI). The first phase of this initiative, completed in 2010, evaluated several facilities that the Corps had identified as requiring rehabilitation. The Corps concluded that modernization of six "critical needs" projects could produce 341 GWH in additional electricity, or an average increase in production of approximately 8% per plant.[29] The cost for these upgrades would be approximately $600 million. The second phase of the HMI looked at facilities outside the Federal Columbia River Power System (i.e., projects not funded by the Bonneville Power Administration, or BPA) and concluded that if the Corps took no action to modernize the 54 units not financed by BPA, it would forego potential revenues of approximately $7 billion over a 20 year horizon.[30] Costs for the upgrades necessary to avoid the aforementioned losses were estimated at $3.7 billion.

Nonfederal Development at Corps Facilities

Besides federal investment and development, nonfederal development of hydropower at Corps sites is also permissible under certain circumstances. According to the Corps, as of 2010 there were 90 nonfederal power units at Corps dams with a total capacity of .003 GW.[31] Such development requires a Federal Energy Regulatory Commission (FERC) license and a Corps Section 408 permit, which authorizes the nonfederal use of a federal facility.[32] The Corps and FERC signed a Memorandum of Understanding (MOU) in March 2011 to coordinate the regulatory review; among other things, the MOU indicates that the Corps will conduct its regulatory review concurrently to the FERC process to the extent possible.[33]

Bureau of Reclamation Projects

Reclamation is the second largest producer of hydroelectric power in the United States. Reclamation operates in the 17 western states, and 11 of these states have Reclamation hydropower facilities.[34] As part of its mission to facilitate settlement of the West through water resources development, Reclamation built numerous projects that included facilities to impound water to provide water supply for irrigation and municipal use and to capture floodwaters. Reclamation also constructed hydropower units at some of these facilities, in part to finance project construction for these other purposes.[35] Reclamation constructed some of these units in the early part of the 20th century, and significantly increased its hydropower production during World War II to meet wartime production demands. Reclamation operates 194 generating units at 58 projects, with total capacity of 14.8 GW, or about 16% of the Nation's hydropower capacity. Reclamation generates on average 40,000 GWH annually.

There are two tiers of customers for federal hydropower projects. For Reclamation projects, power is first used for project purposes (e.g., pumping of water for irrigation); the remaining power is marketed by one of the Power Marketing Administrations with jurisdiction over the area (Bonneville Power Administration (BPA) or the Western Area Power Administration (WAPA). In turn, the PMAs provide this power for distribution and sale (see below section, "Hydropower Receipts and Federal Power Marketing Administrations"). Receipts from hydropower revenues at Reclamation Facilities are deposited into the Reclamation Fund and are first applied to project repayment costs.[36]

Like the Corps, Reclamation has identified challenges associated with the operation of its facilities. Similar to the Corps, Reclamation's facilities are aging: the majority of Reclamation's hydroelectric dams are 40-60 years old, and as of 2007 the average age of its hydropower facilities was 51 years.[37] Reclamation in 2008 estimated that approximately $1.4 billion is needed to upgrade its hydropower assets.[38] However, unlike the Corps, net generation at Reclamation facilities has remained somewhat constant over the last 10 years, and Reclamation has stated that its project performance is generally favorable compared to most industry benchmarks.[39]

Federal Dam Removal and Hydropower

Interest in dam removal for ecosystem and species restoration and recreational purposes is a controversial issue, especially when the dam is producing hydroelectric power. Congress becomes involved as proposals for federal dam removal are considered for authorization and appropriations, as well as in the ongoing implementation of some dam removal projects. There are a few federal facilities with hydropower units where dam removal has been debated. The most controversial case for the Corps is the proposed removal of four Corps locks and dams on the lower Snake River (WA). These facilities have 3GW of capacity (approximately 12% of total Corps capacity) and average annual generation of 7,800 GWH. They also support navigation of the river by commercial barges. The interest in dam removal derives from efforts to restore fish passage for migratory salmon and species listed under the Endangered Species Act. As of 2012, Congress had not acted on proposals for removal of these dams; removal proponents continue to press for these actions.

Reclamation has also studied the potential gains associated with upgrades to its hydropower facilities. However, the potential for additional development at Reclamation facilities appears to be considerably less than that for Corps facilities. In a 2010 study, Reclamation concluded that significant upgrades were only feasible at 10 of its 58 facilities, and would increase Reclamation's total capacity by .067 GW, or less than less than 1%.[40]

Nonfederal Development at Reclamation Facilities

Nonfederal hydropower facilities at existing Reclamation projects is permitted by Reclamation. The process is accomplished through either: 1) a FERC license (for projects not authorized for hydropower); or 2) through a process managed by Reclamation known as the Lease of Power Privilege Process (for projects authorized for hydropower, but not developed). As of 2010, Reclamation reported this type of development at 47 sites which provided capacity of over .46 GW.[41] According to Reclamation, whichever of these processes is used, power development must not conflict with the purposes for which Congress authorized the original Reclamation project, and must not affect the structural and operational integrity of the project. Furthermore, the development must not have significant adverse environmental, cultural, or historical impacts.[42]

Hydropower Receipts and Federal Power Marketing Administrations

The federal government operates four Power Marketing Administrations —Bonneville Power Administration (BPA) in the Pacific Northwest, Southeastern Power Administration (SEPA), Southwestern Power Administration (SWPA), and Western Area Power Administration (WAPA).[43] Each is operated as a distinct and self-contained entity within DOE. Among other responsibilities, the PMAs are responsible for marketing surplus power from Corps and Reclamation facilities to their customers.[44] The revenue collected from the sale of this power is deposited into the Treasury.[45] It is used to pay with interest the federal investment in the hydropower facilities. Gross annual revenue returned to the Treasury from power sales of electricity at federal facilities is estimated at approximately $5 billion, and usually amounts to considerably more than the budgets for hydropower and related operations and maintenance (O&M) for the Corps and Reclamation.

The Flood Control Act of 1944 requires that the PMAs produce power "at the lowest possible rate consistent with sound business practices."[46] This means that the PMAs typically sell power at a lower cost than investor-owned utilities (whose purpose is to provide a financial return to investors). This rate-setting policy was established to encourage regional economic development and appropriate use of federal assets. Some have pointed out that current PMA policies result in long-term, low-cost contracts that do not take into consideration major O&M upgrades and replacement, and are therefore artificially lower than the true cost needed to maintain these facilities.[47]

Federal Hydropower Revenues and Fish Mitigation

Revenue that is collected from hydropower customers is a significant source of the funding for fish mitigation activities in some basins, such as the Columbia River basin. In the mid-2000s, BPA used roughly $260 million annually from hydropower revenues for salmon recovery and mitigation efforts; this represented roughly half of the annual federal funding for Columbia River Endangered Species Act compliance efforts.

Source: Statement of Robert W. Johnson, Commissioner, Bureau of Reclamation, Department of the Interior before the House Committee on Natural Resources, June 12, 2008.

Opportunities and Challenges

Hydropower generation at federal facilities has occurred since the early 20th century and is well established. These projects have provided low-cost power to many, but also have associated environmental costs. As previously noted, several recent studies have noted the potential for increases in federal hydropower, both through efficiency gains and through development of nonfederal power at federal sites. For instance, Corps and Reclamation staff have estimated that approximately 2.4 GW of new capacity (i.e., an increase of as much as 7% on top of existing federal capacity) may be technically feasible to add through additional hydropower development or refurbishment of existing facilities. While fully realizing this potential would likely depend on a number of economic and environmental factors, most agree there is some potential for additional development.

However, challenges to future production and development persist. At Corps facilities in particular, hydroelectric production is facing a number of challenges. A common issue, cited by many, is aging federal hydropower infrastructure, which is often tied to decreasing generation and increases in forced outages of hydropower units. The Federal Government is faced with limited options for dealing with these issues. Financing major upgrades and expansions of federal hydropower facilities beyond immediate maintenance needs is difficult to accomplish without congressional appropriations and, in some cases, authorizations. Construction of new projects also faces challenges. As with expansions and upgrades, new federal developments are dependent upon congressional actions. Other challenges for hydropower at or on federal facilities include

- Environmental operating restrictions in place to protect species that would otherwise be harmed by these projects;
- Uncertain and variable hydrologic conditions (e.g., precipitation and runoff patterns related to climate variability and change);
- Limited operational flexibility (e.g., limits derived from congressional authorization and carefully negotiated operating manuals); and
- Demand for water by other competing uses (e.g., municipal water supply and navigation).

NONFEDERAL HYDROPOWER

Nonfederal hydropower projects can be privately owned or publicly owned, and may or may not be located at a federal site. They differ from federal projects primarily in that they are subject to regulation stemming from their licensing by FERC. According to U.S. Energy Information Administration (EIA) data, roughly 1,245 nonfederal hydropower plants generated 147,400 GWh of net electricity generation in 2010.[48] The amount generated constitutes roughly 57% of total electricity generated from hydropower and roughly 4% of total U.S. electricity in 2010.[49]

In addition to the general advantages and disadvantages of hydropower discussed earlier in this report, nonfederal projects have unique benefits. For example, some argue that nonfederal projects are more likely to receive regular maintenance and major upgrades than federal projects, since the plant owner has a financial incentive to generate as much electricity as possible. On the other hand, drawbacks of nonfederal projects generally mirror those of federal projects (e.g., initial high capital costs, environmental concerns, regulatory requirements). An additional drawback of nonfederal projects is the length of time required to obtain state and federal approval. Some surmise that any increase in hydropower generation is most likely to come from nonfederal projects because, as described earlier, federal projects face hurdles that could prevent them from adding substantial generation capacity quickly, with the major barrier being federal financing for major upgrades and expansions.

Federal Energy Regulatory Commission

FERC licenses the construction and operation of nonfederal hydropower projects.[50] It has the exclusive authority to license new nonfederal hydropower projects, relicense existing projects and provide oversight for all ongoing projects, and it has a role in the decommissioning of projects.[51]

FERC was granted this authority under the Federal Power Act of 1935 (16 U.S.C. 791-828c). FERC may issue a license for a hydroelectric project that

- is located on navigable waters of the United States;
- occupies public lands or reservations of the United States;
- utilizes surplus water or water power from a U.S. government dam; or
- is located on a body of water over which Congress has Commerce Clause jurisdiction, for which project construction occurred on or after August 26, 1935, and which affects interstate or foreign commerce.[52]

FERC reports that the 1,600 projects it regulates at over 2,500 dams represent 54 GW of hydropower capacity, or more than half of all U.S. hydropower.[53] FERC has issued a license for 1,017 projects and has given an exemption for 624 additional projects.[54] Licenses are typically issued for 50 years, but if the relicensing process is underway, an annual license is issued. A 2012 CRS analysis of existing FERC license expiration dates suggests that 47 licensed projects (4.6%) are set to expire between January 2010 and December 2015, 46 licensed projects (4.5%) are set to expire between January 2012 and December 2016, and 90 licensed projects (8.8%) are set to expire between January 2017 and December 2021.[55] No expiration dates are provided by FERC for exemptions, as they are issued in perpetuity.[56]

As an independent agency, FERC's job with respect to hydropower is to license and oversee projects and not necessarily to encourage or discourage additional hydropower development, or to quicken the rate at which projects come online. FERC has worked with stakeholders, as directed by Congress, to streamline its application process. FERC now has three licensing processes that applicants may use to apply for a new license or relicense—the Traditional Licensing Process (TLP), the Alternative Licensing Process (ALP), and the Integrated Licensing Process(ILP).[57] In July 2005, the ILP became the default licensing process (see *Appendix*).[58]

Several factors may affect FERC's ability to keep up with the license application workload. Staff must spend time informing potential applicants about what license they qualify for, and must ensure compliance with various laws. Also, obtaining a FERC license can involve roughly a dozen federal and state agencies, and in some cases the application period may last for six years or more.[59] Federal and state agencies that are consulted during the process may not always adhere to the requisite time frames. For instance, a 2001 FERC report to Congress on hydroelectric license policies notes that a common reason license applications are delayed is untimely receipt of state water quality certification under the Clean Water Act.[60] Testimony by a FERC representative delivered at a 2011 congressional hearing further expresses FERC's perspective:

> Project developers and other stakeholders, not the Commission, in most instances play the leading role in determining project success and whether the regulatory process will be short or long, simple or complex.
> To the extent that a proposed project, even one of small size, raises concerns about water use and other environmental issues, it may be difficult for the Commission to quickly process an application. It is important to remember that the small capacity of a proposed project does not necessarily mean that the project has only minor environmental impacts.[61]

Opportunities and Challenges

In many respects, nonfederal projects face the same challenges and opportunities as federal projects. Both must consider power generation, environmental concerns, and surrounding community issues (e.g., safety, recreation). Furthermore, hydropower projects can come with issues associated with power grid connection and transmission. However, there are some issues unique to nonfederal projects such as satisfying FERC licensing requirements.

Many contend that generation capacity at nonfederal projects could be expanded if the financial incentives were adequate. Congress may consider whether to modify tax incentives or provide other subsidies, and could further examine the role of FERC. Environmental costs and impacts associated with expansion could also be considered.

Challenges unique to nonfederal hydropower projects generally emerge from the regulatory framework. Some regulatory challenges involve the issuance and renewal of licenses in a timely manner, and the adequacy of the federal workforce to oversee the license and exemption process. Other challenges include community opposition, market demand fluctuations, and data submission and analysis to better assess production capability. For some time there has been criticism of FERC's handling of its licensing responsibilities, much of which centers on the assertion that the full potential of nonfederal hydropower is not being delivered because of licensing delays. Some argue that additional generation capacity from hydropower would be possible if FERC's licensing process was less of a barrier, specifically with respect to the amount of information needed to apply for a license and the time it takes to acquire a license. One issue of particular concern to some stakeholders is the requirement that agencies review and provide "mandatory conditions" to protect fish and other resources as part of the licensing process.[62] Others argue that the licensing process needs to be comprehensive, allowing impacted parties adequate time to review the application and offer comment because, once approved, licenses are valid for 30 to 50 years.

Although the challenges for nonfederal projects are significant, there may be opportunities to overcome or minimize the significance of the barriers mentioned above. For example, energy legislation incorporating a federal renewable electricity or clean energy standard that includes hydropower or a price on carbon could absolve some market anxiety and lead to rapid large-scale investments in new nonfederal projects. Moreover, energy efficiency measures and technological advancements could spur additional generation at existing projects.

Other Nonfederal Hydropower Issues

Other pressing issues associated with nonfederal hydropower include its treatment in state renewable portfolio standards, annual charges by FERC for federal lands transferred with a power site classification, and residential and commercial development on the shoreline of FERCregulated hydropower projects. These issues, which are discussed below, may have an indirect impact on the progress of hydropower to meet current and future energy demand.

Role of Hydropower in Renewable Electricity Standards

Some in Congress are interested in strategies that could support more domestic renewable energy production. One proposed strategy is a federal renewable electricity standard (RES) or clean energy standard (CES) (e.g., S. 2146), which could require certain retail electricity suppliers to provide a minimum percentage of the electricity they sell from renewable energy sources or other resources.[63] While a federal RES has yet to be established, many states have created a renewable portfolio standard (RPS)—essentially the same as an RES but carried out at the state level. As of March 2012, there were 29 states and the District of Columbia and Puerto Rico that had an RPS.[64]

The inclusion of hydropower in state standards is not uniform, with each state setting its own criteria. The state standards generally have similar intents, but differ in how they achieve their goals. For example, many have different yearly targets and eligibility requirements for renewable sources. Some state standards include hydropower, but almost always with conditions (e.g., allowing only new hydro projects that have a capacity of 10 MW or less and do not require a new dam, allowing only existing hydro projects of 30 MW or less, excluding pumped storage, only including incremental production at existing facilities). Furthermore, the location and delivery requirements for the electricity generated differs for state standards, with a handful of states having caveats for "geographic eligibility" concerns for hydro projects.[65]

Federal Power Act Section 24 Power Site Reservations

The "power site reservation" under Section 24 of the Federal Power Act has been the subject of a number of administrative disputes over the years.[66] Power site classification is the classification of public lands that have potential value for water power development. Traditionally, the U.S. Geological Survey (USGS) has had the authority to classify the lands as having potential value for water power development (i.e., power site classification), FERC has jurisdiction over the power value on the lands, and BLM has certain management jurisdiction over the surface and subsurface resources, but not the power value. Once land is assigned a power site classification, this classification remains with the land. The classification is not extinguished if the land has been transferred out of federal ownership, although there is a process to extinguish the classification.

An administrative dispute associated with power site classification is that hydropower projects on land transferred from federal ownership to state ownership are still subject to annual fees if the land has a power site classification.[67] According to FERC, annual charges are still required for federal lands that were transferred with a power site classification because the classification remains with the land and the United States still has the right to obtain power from this land.[68] FERC does not keep a record of the number of lands transferred with a power site classification.

Hydroelectric Power and Shoreline Management Plans

Controversies over property rights in areas near hydropower facilities have drawn attention to the FERC "Shoreline Management Plan" (SMP) process. An SMP governs the use and occupancy of the project reservoir shoreline for activities not related to hydropower production. Recent FERC orders proposing to limit acceptable use and occupancy near hydroelectric facilities, including potential interference with existing recreational and other structures, have drawn the ire of nearby communities (see *Appendix*). With an SMP, FERC is trying to ensure that waterfront development along the shoreline of hydropower projects does not have an adverse impact on project operations, public safety, commercial navigation, and other interests. In recent years, property owners along some reservoirs have complained when reservoirs are drawn down to produce power. Such conflicts between shorelines uses can be especially acute during droughts.

The source of FERC's authority to require SMPs and to enforce their terms on hydropower license applicants is primarily found in Sections 10(a)(1) and 4(e) of the FPA.[69] Section 10(a)(1) directs FERC to issue hydropower licenses that include a "comprehensive plan for improving or developing a waterway or waterways for the use or benefit of interstate or foreign commerce, for the improvement and utilization of waterpower development, for the adequate protection mitigation, and enhancement of fish and wildlife ... and for other beneficial public uses." Shoreline management plans are one way that FERC ensures that hydroelectric facilities that it permits satisfy these requirements. The requirement that FERC-permitted hydropower facilities file and abide by SMPs is not in the Federal Power Act (FPA).[70] In fact, the phrase "shoreline management plan" does not appear anywhere in the U.S. Code. The only reference to SMPs is in the titles of the *Code of Federal Regulations* as a requirement that existing licensees include their SMPs in their pre-application filing for renewal of the license.[71]

Within this broad statutory framework, FERC decides on a case-by-case basis whether an SMP is warranted for a proposed hydropower licensee. On occasion an applicant has submitted a proposed SMP without prompting. In some cases, FERC determined an SMP was not needed. Regardless of whether a hydropower licensee is required to submit an SMP, all licensees are required to obtain the property rights needed to satisfy any obligations that may be included in an SMP.

LEGISLATIVE QUESTIONS

Federal Hydropower

What Are Some of the Options to Address Aging Hydropower Infrastructure at Existing Federal Hydropower Projects?

As previously discussed, declining generation trends at federal facilities have been tied to aging infrastructure and the need for funds to replace and/or upgrade this infrastructure. Several policy options have previously been proposed to deal with aging federal hydropower infrastructure. Some of these options are specific to the Corps or Reclamation, while others could generally apply to both agencies. All of these options have budgetary costs in one form or another. Some of the options commonly mentioned include

- *Increase federal funding for hydropower upgrades and/or modernization initiatives within annual appropriations process.* This option would maintain the status quo practice of funding hydropower upgrades within existing discretionary budget allocations, albeit at increased levels or over a different timetable. Some point out that increased funding in annual appropriations is unrealistic in the current budgetary climate.
- *Utilize alternative funding mechanisms to finance hydropower upgrades.* Some have proposed alternative funding mechanisms that fall outside of the regular appropriations process, such as an infrastructure bank or some form of private sector contracting vehicle (such as Energy Savings Performance Contracts, or ESPCs).[72] Advocates argue that these funding mechanisms could increase

hydropower production without requiring annual appropriations. However, utilizing these alternative programs to fund federal projects typically entails "up-front" budgetary costs, especially if they entail federal backing or guarantee future federal spending in the form of project revenues.

- *Increase rates to re-coup "full" costs of operations and maintenance (including major upgrades) and/or institute new user charges to pay for upgrades.* Some argue that rates charged by the PMAs should be increased to recover the costs for major infrastructure upgrades. Congress could alter existing law to allow for these increased rates. However, any practice that results in increased rates for PMA customers may be viewed negatively by customers and some Members of Congress.
- *Privatize federally owned dams through divestiture of assets.* One option sometimes raised is the potential to privatize federally owned hydropower assets, thereby relieving the federal government of its operation and maintenance responsibilities and putting these dams in the hands of other interests, who might better afford to invest in facility upgrades that would increase generation. However, private entities might also increase electricity rates to achieve greater revenues and repay investments in these projects.
- *Allow customers to commit future power revenues to pay for major upgrades to federal facilities.* Some would prefer that those entities who benefit from federal hydropower upgrades be allowed to directly finance these upgrades by redirecting funds that would otherwise flow to the Treasury for these projects. Some PMAs, including BPA, already have some form of this authority, and some advocate for extending it to the other PMAs.[73] There would likely be a budgetary cost to allowing for these receipts to be directed anywhere other than the General Treasury.

What Are Some of the Current Data Gaps Related to Federal Hydropower?

Although a number of the studies previously mentioned in this report collected data on federal hydropower resources, several data gaps have been identified. For instance, one study identified a data gap on the value of ancillary benefits from federal hydropower operations.[74] Ancillary benefits, such as the ability to suddenly increase hydropower production when power is needed (and/or other sources are not available), have the potential to boost the stability and resilience of the nation's power system.[75] However, the role and value that federal hydropower sources currently provide or could provide through ancillary benefits has not been well documented. Additionally, although the Bureau of Reclamation has investigated the economic feasibility of uprating and capacity additions at its facilities, the Corps has not published a similar study. Other gaps related to availability of downscaled climate data of sufficient quality to inform operations of hydropower facilities are a source of ongoing controversy for federal projects.[76]

How Does Uncertainty Affect Operations of Federal Projects?

Uncertainty, in the form of natural climate variability and other related changes, has affected water resources in the past and may affect how much and when water is available for hydroelectric generation in the future. Both the Corps and Reclamation have found that operations of federal hydropower infrastructure will need to be altered due to climate change.[77] Federal facilities are often operated based on procedures set out in operating manuals, which are commonly put together based on past trends and the assumption of "stationarity" (i.e., the idea that future hydrologic trends will be similar to past time periods). Revising these estimates can be difficult, and much uncertainty still exists in the process. Alterations to federal operations are further complicated by existing requirements to meet multiple competing goals in the face of uncertain water conditions. Some stakeholders have questioned whether federal agencies will require greater flexibility in operating existing reservoirs if climate conditions result in changes to weather patterns, less predictability, and more extreme events. In 2009, the Secretary of the Interior, Ken Salazar, issued an order stating that the Department (which has purview over the Bureau of Reclamation) would consider and analyze climate change impacts when undertaking long range planning exercises, but to date the effect of the order on hydropower operations at specific facilities remains unclear.[78]

Nonfederal Hydropower

Should Congress Support Additional Nonfederal Hydropower Projects?

There is a debate about whether more nonfederal hydropower could help to meet current and future electricity demand.[79] Additional hydropower could be less harmful to the environment than fossil-fueled electricity generation projects, if hydropower projects conform to certain standards.[80] Hydropower can also be a flexible source of generation. It can be a baseload energy source, meaning that it is readily available, or it can be used for peaking, meaning that it is turned off and on to meet peak demand. On the other hand, there can be environmental concerns, regulatory concerns, and energy terrorism. Congress could decide whether to support additional hydropower development, how much additional development is necessary, and whether the current regulatory environment for nonfederal hydropower is appropriate. Congress could also determine how involved it wants to be with nonfederal hydropower projects. Hydropower as a whole is one of the few remaining energy sources where the federal government owns a significant portion of the projects that generate the bulk of the electricity. Congress could decide that hydropower development of a certain size and scale is an energy resource that no longer requires such large ownership and intervention by the U.S. government, possibly transferring some of those activities and infrastructure to nonfederal entities.

In What Ways Can Additional Power Be Generated from Existing Nonfederal Projects?

Congress could offer support for additional hydropower generation using numerous mechanisms. Congress could choose to assess current federally funded efforts for nonfederal hydropower projects and take action by providing supplementary funds or modifying efforts to ensure that congressional goals are met on schedule.[81] Congress may or may not expand and extend tax credits available for hydropower (e.g., H.R. 3307). Congress may or may not establish a clean energy standard that includes hydropower as an eligible clean energy source (e.g., S. 2146). Congress may or may not amend specific laws to expedite the licensing process for select projects (e.g., H.R. 2842).

What Can Congress Do to Expedite Nonfederal Project Licensing?

Congress may decide to further examine the role FERC should or should not have in nonfederal hydropower regulation. Reasons may include time delays throughout the licensing process, such as those related to gathering input from multiple affected parties and agencies. Others may argue that FERC's role in nonfederal hydropower regulation should continue for safety reasons and because public resources are being used to generate power. If the length of time required for licensing continues to be a concern, Congress could impress upon federal, state, and local agencies involved in the licensing process the need to complete their contributions in a prompt manner. On the other hand, some stakeholders might object if they believe their concerns (e.g., fish, wildlife, recreation, water use, or other impacts) are not being thoroughly vetted. Legislative proposals that address project licensing include S. 629 which would, in part, direct FERC to investigate the feasibility of issuing a license for certain hydropower projects within a two-year period beginning on the date a prefiling licensing is submitted.

CONCLUSION

If Congress determines that increasing hydropower is a priority, then additional hydropower generation could come from either the public or private sectors or both. It is unclear whether financing to construct, operate, and maintain projects will be a federal priority. The private sector has a sizeable presence in the small hydropower market, which may place the sector in a good position to operate and expand power production. However, there are serious doubts about whether investments in large hydropower projects, by either the public or private sector, are likely to happen soon, owing to economic and geographic constraints,[82] environmental concerns, and public perception. It is possible that other forms of water power (e.g., hydrokinetics, ocean thermal) may one day contribute to the U.S. energy portfolio, but small hydropower is a more likely near-term option.

As Congress considers whether to alter support for hydropower development, it is also dealing with current hydropower infrastructure issues. These issues include aging infrastructure, delayed maintenance, the permitting process, and water availability. Congress may decide to address these concerns in a variety of ways. For example, legislation passed the House in the 112th Congress that would exempt nonfederal projects of 1.5 megawatts or less from certain FERC licensing requirements (i.e., H.R. 2842), and proposed legislation would give tax parity to hydropower projects (e.g., S. 631).[83] Some could argue that hydropower does not need assistance in the form of tax credits because it is an established source and tax credits should be issued to ease the entry of new renewable energy sources to the market or all types of generation should have to compete without federal support. Others could argue that tax credits are needed for hydropower to defray the large capital costs associated with new projects.

Many of the hydropower issues Congress is likely to address in the near term are the same issues it has addressed for some time: operation of federal projects, the permitting process for nonfederal projects, and environmental impacts. The federal government has been responsible for ownership and operation of the bulk of the larger projects. Regular maintenance and upkeep of federal projects has not kept a pace that some would prefer. Also, developers of nonfederal projects would like an easier way to obtain permits and financing for their projects. The nonfederal project permit process has been revised over time in an effort to streamline the process, but still is unacceptable to some.

Hydropower has a long tenure in the electricity market, and its advantages and disadvantages are well documented. However, there are questions for Congress about whether hydropower capacity could or should be increased, and whether that increase should occur at existing projects or by building new projects (including small and low-head hydropower), or both. Other issues affecting conventional hydropower involve the development of non-conventional hydropower technologies, competition from other energy sources (e.g., wind) that are perceived to be more environmentally friendly, and competition from fossil fuel energy sources that may be significantly cheaper (e.g., natural gas).

APPENDIX. FERC INTEGRATED LICENSING PROCESS

FERC's Integrated Licensing Process

The Integrated Licensing Process (ILP) was implemented in 2005 to offer a more efficient licensing process.[84] The ILP incorporates elements of the Traditional Licensing Process (TLP) created in 1985 (e.g., deadlines for multiple steps), and the Alternative Licensing Process (ALP) created in 1997 (e.g., focus on early stakeholder involvement). Additionally, the ILP includes a new process for resolving study disputes and requires FERC to participate earlier in the licensing process. FERC indicates that these changes are intended to make the process shorter and more efficient without altering agencies' authorities under the Federal Power Act (16 U.S.C. §791 et al.) or the Clean Water Act (33 U.S.C. §1341) to develop license conditions that protect fish, federal reservations (e.g., national forests, Indian reservations), or rivers' state-designated uses. The ILP differs from the TLP and the ALP in that it is more collaborative than the TLP and more structured than the ALP. Also, the ILP moves FERC's National Environmental Policy Act (NEPA) scoping process from the post-application phase to the pre-application phase in an effort to resolve study disputes early in the licensing process.

Some entities that might be consulted during the licensing process include:

- National Marine Fisheries Service (NOAA Fisheries)
- U.S. Fish and Wildlife Service (FWS)
- National Park Service (NPS)
- U.S. Environmental Protection Agency (EPA)
- the federal agency administering any United States lands or facilities to be used or occupied by the project
- any state agency with responsibility for fish, wildlife, and botanical resources, water quality, coastal zone management plan consistency certification, shoreline management, and water resources
- the State Historic Preservation Officer (SHPO) and Tribal Historic Preservation Officer (THPO—if applicable)
- local, state, and regional recreation agencies and planning commissions
- local and state zoning agencies
- any Indian tribe that may be affected by the project
- any potentially affected landowners

Applicable laws that might have to be complied with through out the licensing process include:

- Section 401 of the Clean Water Act (CWA)
- Endangered Species Act (ESA)
- Magnuson-Stevens Fishery Conservation and Management Act
- Coastal Zone Management Act (CZMA)
- National Historic Preservation Act (NHPA)
- Pacific Northwest Power Planning and Conservation Act (Act)
- Wild and Scenic Rivers and Wilderness Act

End Notes

[1] This report focuses primarily on conventional hydropower. Conventional hydropower refers to the use of dams or impoundments to store water in a reservoir, whereby water released from the reservoir flows through a turbine to generate electricity. It does not include small hydro, low head hydro, or new hydropower technologies. For more information on these technologies, see CRS Report R41089, Small Hydro and Low-Head Hydro Power Technologies and Prospects, by Richard J. Campbell.

[2] U.S. Energy Information Administration, Electric Power Annual 2010, November 2011, http://www.eia.gov/ electricity/annual/. A gigawatt is equal to one thousand megawatts. EIA defines net generation as the amount of gross generation less the electrical energy consumed at the generating station(s) for station service or auxiliaries. Generator capacity is the maximum output that generating equipment can supply to system load, adjusted for ambient conditions.

[3] The U.S. Department of Energy reports that adding power to nonpowered dams has the potential to add 12 gigawatts of new hydropower capacity, with the 100 largest facilities capable of providing 8 gigawatts. U.S. Department of Energy, An Assessment of Energy Potential at Non-Powered Dams in the United States, April 2012, http://www1.eere. energy.gov/water/pdfs/npd_report.pdf.

[4] Examples of state- and project-specific legislation includes the Bonneville Unit Clean Hydropower Facilitation Act (S. 499) which would authorize the Secretary of the Interior to facilitate the development of hydroelectric power on the Diamond Fork System of the Central Utah Project. Another example is S. 524, which would terminate certain hydropower reservations relating to Bureau of Land Management patents in Madera County, California.

[5] A lease of power privilege is a contractual right given to a nonfederal entity to use a Bureau of Reclamation facility for electric power generation consistent with Reclamation project purposes. The National Environmental Policy Act of 1969 (NEPA, 42 U.S.C. §§4321-4347) requires that a detailed statement of environmental impacts be prepared for all major federal actions significantly affecting the environment. The exemption of certain projects from NEPA compliance has been a contentious aspect of the hydropower debate. For more information on NEPA, see CRS Report RL33152, The National Environmental Policy Act (NEPA): Background and Implementation, by Linda Luther.

[6] Some states do not consider hydropower to be a renewable energy source per state guidelines. For example, California does not consider a new hydropower facility an eligible energy resource for its renewable portfolio standard if it will require a new or increased appropriation or diversion of water under certain Division 2, Part 2 of the California water code..

[7] Certain conditions, such as a drought, which reduce water availability may impact the ability of a hydropower plant to generate the amount of electricity intended. There is an ongoing debate about whether and how to consider potential climate change and its impact on hydropower plant operations during the hydropower plant licensing process. For more information, see Joshua H. Viers, "Hydropower Relicensing and Climate Change," Journal of the American Water Resources Association, August 2011.

[8] M. J. Sale, U.S. Army Corps of Engineers, Institute for Water Resources, Outlook for the U.S. Army Corps of Engineers Hydropower Program, 2011-WRO-P-02, Washington, DC, March 2011, p.1. (Hereinafter referred to as "Corps Hydropower Outlook").

[9] More information on the types of hydropower plants is available at http://www1.eere. energy.gov/water/ hydro_plant_types.html.

[10] A penstock is a closed conduit or pipe for conducting water to the powerhouse.

[11] At one time, hydropower accounted for 40% of U.S. electricity generation. Over time, as other electricity sources expanded, hydropower's share decreased. Other reasons for the decline in electricity generation from hydropower include aging infrastructure, market conditions, environmental and ecosystem concerns, and the fact that large conventional hydropower projects are mostly already developed.

[12] U.S. Energy Information Administration, Hydropower Has a Long History in the United States, June 8, 2011, http://www.eia.gov/todayinenergy/detail.cfm?id=2130.

[13] U.S. Energy Information Administration, Electric Power Monthly, October 2011, http://www.eia.gov/electricity/ monthly/; U.S. Energy Information Administration, Electric Power Monthly, December 2011, http://www.eia.doe.gov/ cneaf/electricity/epm/ epm_sum.html. EIA defines net generation as the amount of gross generation less the electrical energy consumed at the generating station(s) for station service or auxiliaries. Electricity required for pumping at pumped-storage plants is regarded as electricity for station service and is deducted from gross generation.

[14] U.S. Energy Information Administration, Electricity Data, February 2012, http://www.eia.gov/electricity/ data.cfm#generation.

[15] U.S. Energy Information Administration, Electric Power Monthly, February 2012.

[16] This is especially true for nonfederal hydropower projects, as one project could have multiple generating stations.

[17] U.S. Army Corps of Engineers , U.S. Army Corps of Engineers Civil Works, 2009, http://www.usace.army.mil/ CECW/Documents/cecwm/5_yr1/fy09_5yrplan.pdf; Bureau of Reclamation, Hydroelectric Powerplants Operated by Others, 2011, http://www.usbr.gov/ power/data/faclothr.html.

[18] Douglas G. Hall and Kelly S. Reeves, A Study of United States Hydroelectric Plant Ownership, Idaho National Laboratory, June 2006, http://hydropower.inel. gov/hydrofacts/ pdfs/ a_study_of_united_states_hydroelectric_plant_ownership.pdf. The study categorized hydropower plant ownership in six classes: federal, industrial, cooperative, private utility, private non-utility, and municipal and other nonfederal public.

[19] Ibid.

[20] Douglas G. Hall and Kelly S. Reeves, A Study of United States Hydroelectric Plant Ownership, Idaho National Laboratory, INL/EXT-06-11519, Idaho Falls, ID, June 2006, (hereinafter referred to as "Hall and Reeves"). http://www.inl.gov/technicalpublications/ Documents/3374828.pdf.

[21] Hall and Reeves, p vi.

[22] Corps Hydropower Outlook, p. 9.

[23] U.S. Congress, House Committee on Natural Resources, Subcommittee on Water and Power, Statement of Michael Ensch, Chief of Operations, U.S. Army Corps of Engineers, Investment in Small Hydropower: Prospects of Expanding Low-Impact and Affordable Hydropower Generation in the West, 111th Cong., 2nd sess., July 29, 2010, p. 2. Hereinafter referred to as "Ensch."

[24] Corps Hydropower Outlook, p. 2.

[25] Corps Hydropower Outlook, p. 9. Availability of water, including the lack of water in certain hydrological conditions and instream flow requirements, are some of the other factors that may account for decreased generation.

[26] CRS analysis of Corps and EIA data. Again, this calculation does not account for water availability in different regions of the country, which has the ability to effect the Corps contribution to overall generation.

[27] "Unit availability" is generally defined as the amount of time (or in this case, percentage of time) a unit is available to produce electricity, regardless of whether or not it is operated.

[28] Corps Hydropower Outlook, p. 10.

[29] Ensch, p. 3.

[30] Montgomery, Watson, Harza, Phase II Needs and Opportunities Evaluation and Ranking, Reconnaissance Hydroelectric Assessment Plants Life Extensions and Upgrades,

Cumberland River Basin. Report prepared for the U.S. Army Corps of Engineers Nashville District, Contract No. W912P5-06-D-0008-0004.

[31] Ensch, p. 3.

[32] 33 U.S.C. §408.

[33] Corps and FERC, Memorandum of Understanding between United States Army Corps of Engineers and the Federal Energy Regulatory Commission on Non-Federal Hydropower Projects, http://www.ferc.gov/legal/maj-ord-reg/mou/ mou-usace.pdf.

[34] Reclamation operates hydropower projects in the following states (in order of generation capacity): Washington, Oregon, California, Arizona, Nevada, Idaho, Montana, Wyoming, Utah, Colorado, and New Mexico.

[35] Although the Reclamation Act of 1902 was the original statute authorizing construction of Reclamation projects. The Reclamation Service (later renamed the Bureau of Reclamation) was first authorized to develop hydropower resources in the Town Site and Power Development Act of 1906 (34 Stat 116). The Fact Finders' Act of 1924 (43 Stat 701) later authorized the use of hydropower revenues as a credit to construction charges of a project.

[36] For more on the Reclamation Fund, see CRS Report R41844, The Reclamation Fund: A Primer, by Charles V. Stern.

[37] U.S. Bureau of Reclamation, Reclamation-Wide Power Profile. Available at http://www.usbr.gov/power/data/reclwid.pdf.

[38] Statement of Robert W. Johnson, U.S. Congress, Senate Committee on Energy and Natural Resources, Subcommittee on Water and Power, Statement of Robert W. Johnson, Hearing on Reclamation's Aging 110th Cong., 2nd sess., April 2008.

[39] This statement takes into account "available water" for generation, which Reclamation tracks. For instance, see http://www.usbr.gov/power/data/recl-wid.pdf, p. 6.

[40] Bureau of Reclamation, Assessment of Potential Capacity Increases at Existing Hydropower Plants, October 2010, http://www.usbr.gov/power/AssessmentReport/ USBRHMICapacity AdditionFinalReportOctober2010.pdf.

[41] U.S. Congress, House Committee on Natural Resources, Subcommittee on Water and Power, Statement Michael Conner, Commission, Bureau of Reclamation, Hearing on Investment in Small Hydropower: Prospects of Expanding Low-Impact and Affordable Hydropower Generation in the West, 111th Cong., 2nd sess., July 29, 2010.

[42] For more information on the program, see http://www.usbr.gov/uc/power/ progact/ nonfedpwr.html.

[43] For more information, see CRS Report R41960, Federal Agency Authority to Contract for Electric Power and Renewable Energy Supply, by Anthony Andrews.

[44] Some of the PMAs also own and operate transmission lines associated with these facilities.

[45] Revenues from most hydropower receipts are deposited into the General Fund of the Treasury, although as previously discussed, some hydropower receipts at Reclamation projects are credited to the Reclamation Fund.

[46] 16 U.S.C. §825s.

[47] Corps Hydropower Outlook, p. 39.

[48] Energy Information Administration, 2010 December EIA-923 Monthly Time Series File, http://www.eia.gov/cneaf/ electricity/page/eia906_920.html. The amount generated is enough to power roughly 13 million "average" homes, based on EIA 2010 data for the average annual electricity consumption (11,496 kWh, or 11.5 MWh)) for a U.S. residential utility customer.

[49] CRS calculations do not include pumped storage.

[50] Three divisions under the FERC Office of Energy Projects (OEP) administer the hydropower responsibilities of the Commission—the Division of Hydropower Licensing (DHL), the Division of Hydropower Administration and Compliance (DHAC), and the Division of Dam Safety and Compliance (DDSC). DHL is responsible for case management and order preparation for applications for licenses, relicensing, exemptions, major amendment of licenses, 5MW exemptions, and license surrender of constructed projects. DHL is also responsible for the preparation of NEPA documents (e.g., environmental assessments and environmental impact statements) and pre-filing collaborative work. DHAC is primarily responsible for reviewing and ensuring compliance by owners of hydropower projects with the conditions specified in their licenses and exemptions. DDSC is responsible for construction, operation, exemption, prelicense, and environmental and public use inspections; engineering evaluations and studies; independent consultant report reviews; emergency action plan development and testing; engineering guidelines development; and interagency/industry committee participation.

[51] In addition to issuing licenses, FERC reports that it has the legal authority to deny a new license at the time of relicensing. For more on license denial, dam removal, and dam decommissioning, see CRS Report RL33480, Dam Removal: Issues, Considerations, and Controversies, by Nic Lane; and Federal Energy Regulatory Commission, "Project Decommissioning at Relicensing; Policy Statement," 60 Federal Register 339-356, Jan. 4, 1995.

[52] 16 U.S.C §817.

[53] Testimony of Jeff C. Wright, in U.S. Congress, Senate Committee on Energy and Natural Resources, Full Committee Hearing: To Hear Testimony on S. 629, S. 630, and Title I, Subtitle D of the American Clean Energy Leadership Act of 2009, 111th Cong., 1st sess., March 31, 2011.

[54] A list of licenses and exemptions is available at the FERC Hydropower website: http://www.ferc.gov/industries/ hydropower.asp. Reasons for the difference in the number of nonfederal facilities (roughly 1,600 according to FERC and roughly 1,245 according to EIA) are unclear.

[55] According to a Nov. 2010 letter submitted to Congress by FERC, licenses for over 50 hydroelectric projects have expired or will expire between January 2010 and December 2015. See http://www.ferc.gov/about/strat-docs/fy09-anrpt.pdf.

[56] Some projects are granted an exemption which exempts the project from licensing. FERC offers two types of exemptions: small hydropower and conduits. For more information on exemptions, see http://www.ferc.gov/industries/ hydropower/gen-info/licensing/ exemp tions.asp.

[57] For background on licensing processes, see CRS Report RL31903, Relicensing of Nonfederal Hydroelectric Projects: Background and Procedural Reform Issues, by Nic Lane.

[58] Applicants may petition to use either the ALP or the TLP based on anticipated costs, level of complexity and controversy related to licensing, and other factors.

[59] An initial consultation contact list is available for each state, with information on federal, state, and regional agencies and more, at http://www.ferc.gov/industries/hydropower/ enviro/consultlist.aspx. For an account of the changes to hydropower license processes over time, see CRS Report RL31903, Relicensing of Nonfederal Hydroelectric Projects: Background and Procedural Reform Issues, by Nic Lane.

[60] FERC, Report on Hydroelectric Licensing Policies, Procedures, and Regulations Comprehensive Review and Recommendations Pursuant to Section 603 of the Energy Act of 2000, May 2001, http://www.ferc.gov/legal/maj-ordreg/land-docs/ortc_final.pdf.

[61] Testimony of Jeff C. Wright, in U.S. Congress, Senate Committee on Energy and Natural Resources, Full Committee Hearing: to hear testimony on S. 629, S. 630, and Title I, Subtitle D of the American Clean Energy Leadership Act of 2009, 111th Cong., 1st sess., March 31, 2011.

[62] For more information on the licensing process, see Appendix. For more information on mandatory conditions, see Federal Energy Regulatory Commission, U.S. Department of Interior, and U.S. Department of Commerce, et al., Interagency Task Force Report on Agency Recommendations, Conditions, and Prescriptions Under Part I of the Federal Power Act, December 2000, http://www.ferc.gov/industries/hydropower/indus-act/itf/fpa_ final.pdf.

[63] For more information, see CRS Report R42522, Clean Energy Standard: Summary and Analysis of S. 2146, by Phillip Brown, CRS Report R41797, Clean Energy Standard: Potential Qualifying Energy Sources, coordinated by Kelsi Bracmort, and CRS Report R41493, Options for a Federal Renewable Electricity Standard, by Richard J. Campbell.

[64] DSIRE, Quantitative RPS Data Project, October 2011, http://www.dsireusa.org/rpsdata/index.cfm.

[65] Union of Concerned Scientists, Renewable Electricity Standards Toolkit, http://go.ucsusa.org/cgi-bin/RES/ state_standards_search.pl?template=main.

[66] Section 24 (16 U.S.C. §818) establishes that when FERC licenses a hydropower facility on federal land, it reserves lands associated with that facility for the federal government, and that if and when such lands are later transferred by the federal government, the government reserves the ability to continue to regulate the hydroelectric facility.

[67] In November 2011, FERC issued a proposal to revise the methodology for assessing annual charges for the use of federal lands by hydropower projects. Federal Energy Regulatory Commission, "Annual Charges for Use of Government Lands," 76 Federal Register 72134-72142, November 22, 2011.

[68] Personal communication with FERC Office of the General Counsel, January 31, 2012. Annual charges for hydropower projects allow FERC to recover costs incurred in the performance of its regulatory responsibilities. Information on annual charges, including FERC's proposal to revise the methodology for calculating rental rates for the use of federal lands by hydropower projects is available at http://www.ferc.gov/industries/hydropower/annual-charges.asp.

[69] 16 U.S.C. §803(a)(1).

[70] 16 U.S.C. §792 et seq.

[71] 18 C.F.R. §5.6.

[72] More information on ESPCs is available at http://www1.eere.energy.gov/ femp/financing/espcs.html. Currently, federal hydroelectric projects are ineligible for ESPCs.

[73] Specifically, Congress provided BPA with "self-financing" authority in 1974 (P.L. 93-454) by establishing a separate fund in the Treasury, the Bonneville Fund, in which BPA deposits and manages its revenues. The fund allows for BPA to enter into multi-year commitments. BPA also has the authority to borrow from the Treasury; it must repay these loans with interest at market rates.

[74] Corps Hydropower Outlook, p. 51.

[75] Another ancillary benefit is hydropower's ability to restart electric power systems in the event of a blackout; this is known as "black start."

[76] Joshua H. Viers, "Hydropower Relicensing and Climate Change," Journal of the American Water Resources Association, vol. 47, no. 4 (August 2011), pp. 655-661.

[77] For example, see the Corps climate response page at http://corpsclimate.us/responses.cfm. Reclamation research on climate change is compiled at http://www.usbr.gov/climate/ SECURE/.

[78] Department of the Interior, Addressing the Impacts of Climate Change on America's Water, Land, and Other Natural Resources, Order No. 3289, Washington, DC, September 14, 2009, http://www.doi.gov/whatwedo/climate/cop15/ upload/SecOrder3289.pdf.

[79] For a variety of reasons, electricity generation from coal is predicted to be on the decline. The energy source that will make up for lost electricity generation from coal is currently under discussion, but most new capacity additions are expected to be from natural gas generators. Any one energy source or mixture of sources could also be tapped to meet consumer demand.

[80] While hydropower is often viewed as being a clean and renewable source of energy, the environmental costs due to stream flow disruption, particularly effects on fish and wildlife, can be quite high. For example, the Bonneville Power Administration and Reclamation spend hundreds of millions on salmon restoration and mitigation activities associated with federal dams and hydropower facilities. Requiring fish mitigation strategies upfront might avoid costly habitat mitigation later.

[81] For example, DOE supports conventional hydropower through its water power research and development program. The program's conventional hydropower activities are focused on new technologies, optimization, and siting techniques that might improve combined energy and environmental performance.

[82] Most of the large, economically feasible generation sites were developed in the 20th century.

[83] Other electricity sources receive tax incentives.

[84] FERC receives an annual appropriation from Congress to defray its operating costs and recovers 100 percent of this appropriation through the collection of annual charges and filing fees. There are no filing fees for hydropower projects. However, there are annual charges for hydropower projects if the project capacity is greater than 1.5 megawatts (MW). The annual charges for hydropower projects in 2010 totaled nearly $81 million (The total included 2010 actual administrative charges, 2010 actual fix annual charges, and 2010 other federal agency annual charges. http://www.ferc.gov/industries/hydropower/annual-charges. asp).

In: Hydropower and Energy Potential ... ISBN 978-1-62257-711-8
Editors: K. F. Anderson, S. A. Collins

Chapter 2

AN ASSESSMENT OF ENERGY POTENTIAL AT NON-POWERED DAMS IN THE UNITED STATES*

Boualem Hadjerioua, Yaxing Wei and Shih-Chieh Kao

EXECUTIVE SUMMARY

The United States has produced clean, renewable electricity from hydropower for more than 100 years, but hydropower-producing facilities represent only a fraction of the infrastructure development that has taken place on the nation's waterways. In contrast to the roughly 2,500 dams that provide 78 gigawatts (GW)[1] of conventional and 22 GW of pumped-storage hydropower, the United States has more than 80,000 non-powered dams (NPDs)—dams that do not produce electricity—providing a variety of services ranging from water supply to inland navigation. Importantly, many of the monetary costs and environmental impacts of dam construction have already been incurred at NPDs, so adding power to the existing dam structure can often be achieved at lower cost, with less risk, and in a shorter timeframe than development requiring new dam construction. The abundance, cost, and environmental favorability of NPDs, combined with the reliability and

* This is an edited, reformatted and augmented version of a U.S. Department of Energy publication, Prepared by Oak Ridge National Laboratory.

predictability of hydropower, make these dams a highly attractive source for expanding the nation's renewable energy supply.

To better characterize this unique national resource, the U.S. Department of Energy (DOE) Wind and Water Power Program has undertaken a national-scale analysis of U.S. dams to determine the ability of NPDs to provide hydroelectric power. DOE's Oak Ridge National Laboratory (ORNL), with input from DOE's Idaho National Laboratory, quantified the potential capacity and generation available from adding power production capability to U.S. NPDs. Of the more than 80,000 NPDs throughout the U.S., 54,391 dams were analyzed, with remaining dams eliminated from consideration due to erroneous geographic information, or erroneous flow or drainage area attributes that could not be resolved and corrected through independent investigation of maps and records. Anecdotal information suggests that these dams with missing or erroneous information are likely to be relatively small or have low potential to produce hydroelectric energy. Dams with a reported height of less than five feet were also excluded from analysis. A thorough quality control and review process ensured that the 54,391 remaining NPDs were analyzed and characterized as accurately as possible. Figure ES-1 demonstrates the spatial and capacity potential distribution of the nation's NPDs. Electric generating capacities included in the report were calculated using the assumption that all water passing a facility would be available for conversion into electrical energy and that hydraulic head at the facility would remain constant. The analysis did not consider the economic feasibility of developing each unpowered facility. The assessment provides preliminary information for stakeholders (such as developers, municipal planners, and policymakers), who can further evaluate the potential to increase hydropower production at NPD sites. Developers could use the information provided in this assessment to focus on more detailed analysis of sites that demonstrate a reasonable potential for being developed.

Adding power to U.S. NPDs has the potential to add up to 12 GW (12,000 megawatts or MW) of new renewable capacity—a potential equivalent to increasing the size of the existing conventional hydropower fleet by 15%. A majority of this potential is concentrated in just 100 NPDs, which could contribute approximately 8 GW of clean, reliable hydropower; the top 10 facilities alone could add up to 3 GW of new hydropower. Eighty-one of the 100 top NPDs are U.S. Army Corps of Engineers (USACE) facilities, many of which, including all of the top 10, are navigation locks on the Ohio River, Mississippi River, Alabama River, and Arkansas River, as well as their major tributaries. This study also shows that dams owned by the U.S. Bureau of

Reclamation hold the potential to add approximately 260 MW of capacity; the Bureau has also engaged in an effort to conduct a more detailed evaluation of its own facilities.

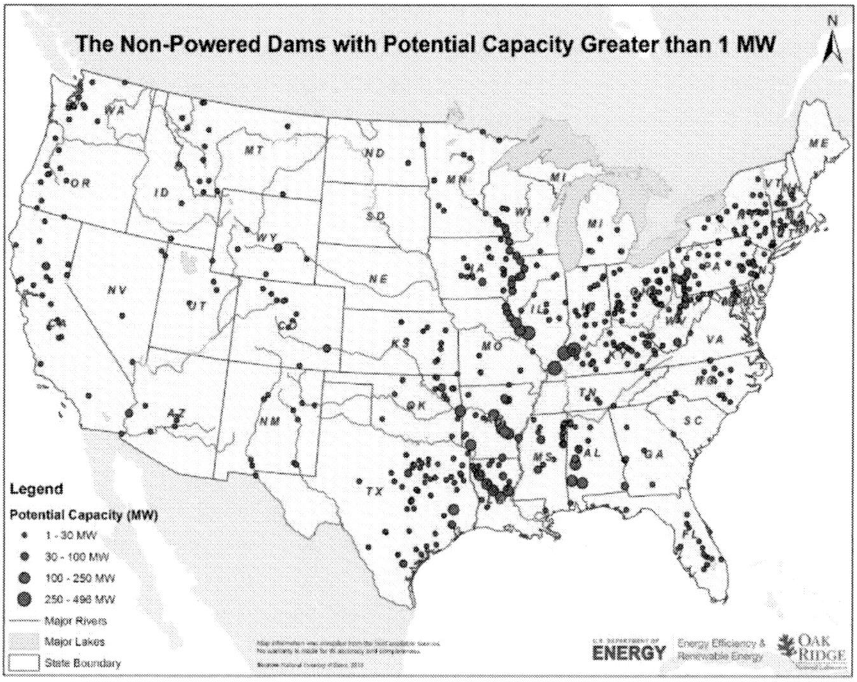

Figure ES-1. Locations of the top non-powered dams with potential hydropower capacities greater than 1 MW.

1. INTRODUCTION

This report describes a U.S. Department of Energy (DOE) Wind and Water Power Program study carried out by Oak Ridge National Laboratory (ORNL) with inputs from Idaho National Laboratory to assess the energy potential at non-powered dams (NPDs) throughout the United States. In this context, NPDs are dams that do not include hydraulic turbine (hydropower) equipment. Such dams were constructed for one or more non-energy benefits, including flood control, water supply, navigation, or recreation. Auxiliary dams that form parts of power-producing impoundments are not included in the NPD scope because the water they store already is associated with existing hydropower production.

This report addresses only the energy production potential of NPDs. The priority placed on this NPD assessment effort (relative to an assessment of energy potential from new impoundments, for example) is based on the hypothesis that many of the costs and environmental impacts of dam construction have already been incurred at NPDs and may not be significantly increased by the incorporation of new energy production facilities. Thus, the development of some NPD's for energy purposes is assumed to be achievable with lower installed cost, lower levelized cost-of-energy, fewer barriers to development, less technological and business risk, and in a shorter time frame than development requiring new dam construction. Future detailed studies of site-specific costs and impacts will be required to test this hypothesis. That is, the initial estimates of the number of developable NPDs and associated capacity additions will have to be refined with better information on such issues as environmental constraints (e.g., the environmental costs and benefits of changes in flow releases to optimize power production), dam integrity/ safety issues, and multiple use conflicts. NPDs in the United States range in size from small berms impounding farm ponds to large Ohio River and Mississippi River dams that pool water to maintain navigation depths during low-flow periods. The National Inventory of Dams (NID) includes more than 80,000 dams with physical heights ranging from about 4 feet to 770 feet. This study analyzed a subset of 54,391 NPDs with monthly average flows ranging from about 1 cubic feet per second (cfs) to 68,500 cfs.

1.1. The Context for NPD Resource Assessment

The assessment activities reported herein were aimed at quantifying, with a useful degree of certainty, the aggregated national, regional, and basin-scale added capacity and generation potential for NPDs. NPDs are one of several resource classes to be examined in the DOE Hydropower Resource Assessment. Additional resource classes include the following:

- Increased capacity and generation at existing powered facilities through unit upgrades, minimum flows units or additional units.
- New capacity from constructed waterways, such as irrigation canals and municipal water systems.
- New capacity and storage from new sustainable hydropower sites.

Each site considered for upgrade or new development in any of these resource classes has a set of cost, socioeconomic impact, and environmental impact attributes that have influenced and will influence the hydropower development and regulatory decisions for that site. This initial report on NPD potential does not address these costs and impacts. Development also is influenced by energy prices in different regions of the United States along with other incentives, such as state programs that provide tax credits or renewable energy credits. At national and regional scales (the focus of this initial report), the geospatial databases included in the assessments reported herein enable automated analyses and aggregation of site potentials according to explicit rules. These rules include geo-registration to assign a dam to a specific stream segment, estimation of seasonal and monthly water availability, estimation of hydraulic head, and estimation of a capacity factor to determine rated capacity from annual production. Such rules engender assumptions, approximations, and uncertainty that render results that are only appropriate for reconnaissance -level studies of hydropower development.

In contrast to site-specific studies that would support project design, feasibility, and due-diligence studies for project development financing, the focus of these national- and regional-scale studies is to provide high-level summary statistics of the availability of hydropower resources. Avoidance of systematic processing errors that overestimate or underestimate aggregate national and regional energy and capacity statistics is a key concern. Conversely, uncertainty in the capacity and energy production estimates for individual sites (i.e., from relying on remote-sensing data and maps rather than "boots on the ground" observations of conditions) is problematic only if it biases trends among technology classes or geographic regions.

The availability and enhancement of national-scale datasets, such as NID and the National Hydrography Dataset (NHD), provide a foundation for rigorous estimates of NPD potential for production for the entire country. Previous assessments—including Hall et al. (2004), U.S. Department of Interior (DOI) et al. (2007), and U.S. Bureau of Reclamation (2011)—have also been consulted in preparing the estimates reported herein.

1.2. Water Availability and Hydrologic Regions

Water availability, along with physical relief, is the primary determinant of energy potential. It varies significantly among different geographical regions in the United States. To help understand our current resources at a regional level, Table 1 summarizes data for two important hydrologic variables, precipitation (P) and runoff (Q), and the Q/P ratios for the years 1971–2008 for each of the 18 major hydrologic regions of the continental United States. Both P and Q are expressed in inches. The Q/P ratio indicates the percent of precipitation that can eventually be utilized from streamflow after evapotranspiration, groundwater infiltration, and other internal hydrologic processes. Evaporation is believed to be the dominant factor affecting the Q/P ratio. Although the Q/P ratio does not directly contribute to the assessment of NPD potential, it can provide some background hydrologic understanding regarding the regional water availability that may be utilized for hydropower generation. For comparison, the existing hydropower capacity within each hydrologic region is illustrated in Figure 1.

Table 1. Mean Annual U.S. Water Availability from 1971–2008

Hydrologic Regions (HUC02)	Precip. P[a] (inch)	Runoff Q[b] (inch)	Q/P ratio (%)	Hydrologic Regions (HUC02)	Precip. P[a] (inch)	Runoff Q[b] (inch)	Q/P ratio (%)
1 New England	45.7	25.7	56.1	10 Missouri	20.7	2.5	11.8
2 Mid Atlantic	44.2	20.3	46.1	11 Arkansas-White-Red	31.7	2.9	9.1
3 South Atlantic-Gulf	52.1	17.2	33.1	12 Texas-Gulf	32.6	2.3	7.0
4 Great Lakes	34.5	14.0	40.5	13 Rio Grande	14.9	0.5	3.5
5 Ohio	45.1	19.4	43.0	14 Upper Colorado	14.9	1.8	11.9
6 Tennessee	54.6	23.2	42.5	15 Lower Colorado	12.7	0.4	3.5
7 Upper Mississippi	33.9	9.8	29.0	16 Great Basin	12.3	2.0	16.6
8 Lower Mississippi	56.5	19.1	33.9	17 Pacific Northwest	33.4	13.5	40.4
9 Souris-Red-Rainy	21.4	2.6	12.1	18 California	24.0	10.5	43.6

[a] Precipitation data was obtained from the PRISM Research Group, Oregon State University.

[b] Runoff data was obtained from the U.S. Geological Survey (USGS) WaterWatch Program.

For the Northeast and Great Lakes Regions (Regions 1, 2, and 4), both precipitation and runoff were abundant. The Q/P ratios are high due to higher latitude, since cooler temperatures reduce evaporation and evapotranspiration. Due to the sufficient water resources, there has been extensive, mostly non-federal hydropower development in these regions. For the South Atlantic-Gulf Region (Region 3), the precipitation is among the highest, mostly due to hurricanes and summer convective storms. Although the high evaporation reduces the available runoff, there are still abundant water resources for hydropower generation. The high summer precipitation brings in challenges related to flood operation. The overall Mississippi River Basin (Regions

5–8, 10, and 11) covers nearly half of the U.S. territory. The eastern areas—the Ohio and Tennessee Regions (Regions 5 and 6)—are the wettest with low evaporation. These two regions have been historically important for hydropower generation, and they are still good potentials for future development. The western parts of the Mississippi River Basin (Regions 10 and 11) are much drier, and the greater evaporation causes large losses in effective precipitation. Other competing, consumptive water uses (e.g., municipal and irrigation) become important and may restrict the water availability for hydropower generation. The huge watershed area of the Mississippi River system results in high magnitude of streamflow downstream (Regions 7 and 8). To assist the river transportation, a series of locks and dams were built by the United States Army Corps of Engineers (USACE) to control the river stage. Many of these dams have not been utilized for hydropower generation, and—it will be shown later—they have the highest potential for NPD development.

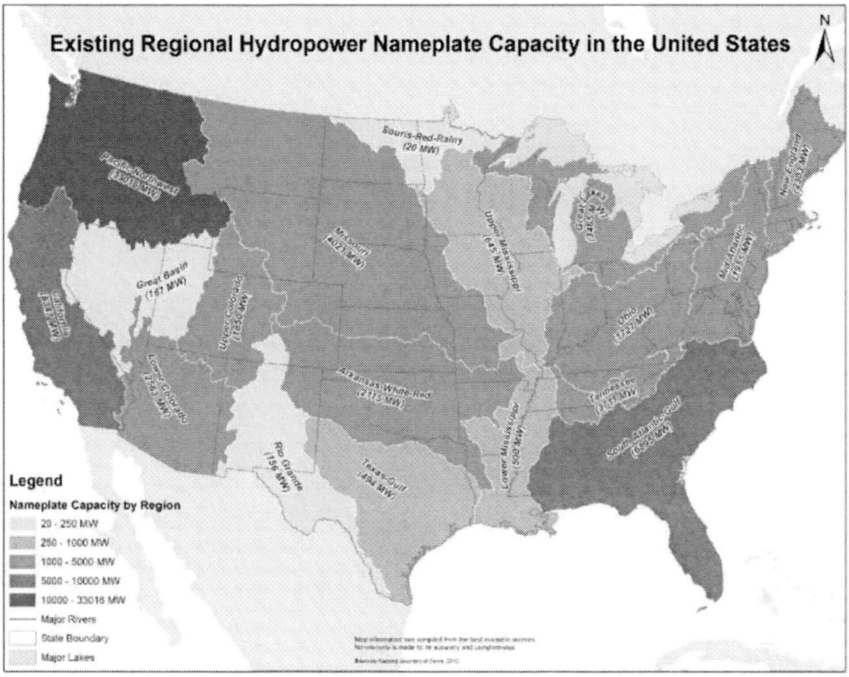

Figure 1. The existing hydropower capacity within each hydrologic region (Source: National Hydropower Asset Assessment Program Existing Hydropower Plant Summary, Hadjerioua et al., 2011).

The Souris-Red-Rainy Region (Region 9) is the upstream portion of river systems that drain north into Canada. Due to lower populations and available water resources, hydropower has not been a focus in this region. The Texas-Gulf and Rio Grande Regions (Regions 12 and 13) are among the driest in the United States.

Although most of the NPDs are located in the State of Texas, the future hydropower potential is, in fact, limited by available precipitation and low Q/P ratios. The Colorado River system (Regions 14 and 15) also has relatively low precipitation and runoff. The regional water supply heavily relies on storage in large reservoirs, such as Glen Canyon Dam and Hoover Dam. Although not the main focus, these huge water storage projects permit high hydropower generation in these regions. The Great Basin (Region 16) is a closed watershed with limited water resources; there are comparatively fewer existing hydropower projects in this region.

The Pacific Northwest (Region 17) is the most important region for U.S. hydropower generation. Both precipitation and runoff are abundant, and evaporation losses are low. The largest U.S. hydropower plant, Grand Coulee, is located in this region along with several other large projects (e.g., Chief Joseph). California (Region 18) is also an important hydropower provider; the numerous Central Valley projects in California create a large capacity for hydropower generation. One special feature of California is that flow diversion was utilized in several projects, resulting in high hydraulic head for hydropower generation.

Although both the Pacific Northwest and California already provide large amounts of hydropower, the environmental concerns, such as fish habitat, may pose challenges for future development. Because the western United States (Regions 10–18) is much drier overall than the eastern regions, hydropower generation needs to be coordinated with other competing water usage. The DOI's Bureau of Reclamation (Reclamation) is assigned the responsibility of water management and supply in the western United States.

2. METHODOLOGY

2.1. Energy Production Models for a Non-Powered Dam

Consistent with previous studies (DOI et al., 2007; Reclamation, 2011), the following formula was utilized to estimate the potential hydropower generation at NPD sites:

Potential Hydropower Generation (in megawatt hours or MWh) =
$Q * \Delta H * \eta * T / 11800$ q (1)

In Equation (1), ΔH (ft) is the gross head for hydropower generation, η is the generating efficiency (currently assumed to be 0.85 in this assessment), and Q (ft^3/s) is the average flow during the total generation period T (hour). While Equation (1) provides a straightforward way to estimate the power potential, there remain true challenges regarding how to estimate the parameters for a large number of NPDs given the available data. Ideally, the flow-duration curve based on at least 10 years of record needs to be developed at each NPD site so that wet, dry, and normal hydrological years can be captured. However, such a task is challenging at the national scale since most of the NPDs are located on ungauged streams or have limited historical observation. Therefore, in this assessment Q cannot be estimated by the commonly used 30% exceedance level from daily flow-duration curves (criterion used in Reclamation, 2011). Estimating the gross head—ΔH— available at the NPD with accuracy also is challenging. It is not possible, within the present effort, to obtain and analyze a daily headwater and tailwater elevation time series for each NPD. Thus, the potential energy must be estimated from available statistics for each site. For a national-level analysis, reasonable alternatives must be sought.

In this assessment, monthly hydropower generation was computed using a constant gross head derived from the NID and other methods (see head adjustment, Chapter 2.4) and the estimated monthly mean flow at each NPD site assuming that all flow can be utilized for hydropower generation. By summing the hydropower generation for each month, the potential annual generation can be estimated; this represents the maximum theoretical hydropower generation one may obtain at an NPD site. A regional capacity factor (C_f), which was computed from existing hydropower plants (Hadjerioua et al., 2011), was then utilized to estimate the potential capacity.

- *Capacity Factor Definition:* The ratio of the actual energy produced in a given period, to the hypothetical maximum possible—i.e., running full-time at rated power.
- *Regional Computed Capacity Factor:* Capacity factor for this study was computed based on 8-year (2001–2008) recorded Energy Information Administration (EIA) generation for all hydropower plants in the United States; a regional capacity factor then was

developed and assigned to compute the potential site installed capacity.

- *Capacity Factor Limitations:* Since the actual hydropower production is mainly a function of hydrology, operational constraints, demand, and dam type (tributary, main stem, run of river, etc.), there will be some overestimation or underestimation in the computed installed capacity when using an overall regional historical capacity factor. It is expected that appropriate site-specific capacity factors could be determined after a detailed feasibility study addressing all site characteristics (hydrology, hydraulic, environment, demand, machinery, etc.).

The capacity factor was computed by:

$$\text{Capacity Factor (Cf)} = \text{Annual Generation}/(\text{Installed Capacity} * 365 * 24) \quad (2)$$

In Equation (2), the actual historical generation was observed at existing hydropower plants (collected from EIA, 2010). If all generating units in a hydropower plant run continuously year-long, the actual generation will be close to the installed capacity multiplied by the entire hours in one year; hence, the capacity factor will be close to one. However, because streamflow fluctuates significantly over time, actual capacity factors are significantly less than one. The Cf computed from existing plants provides one way to estimate the installed capacity from Equation (1). The following energy production model is therefore utilized in this assessment:

$$\text{Potential Capacity (MW)} = \text{Potential Power Generation (MWh)}/(\text{Cf} * 365 * 24) \quad (3)$$

By using the above equations jointly, the potential hydropower generation and capacity are estimated at each NPD site. However, this capacity value assumes construction of a powerhouse that can pass all available water for the site for the generation of electricity; designed and developed powerhouses subject to economic limitations on size will be significantly smaller in capacity.

Several national-scale datasets (listed in Figure 2) were utilized for the assessment, and each component is discussed in the following subsections. For the reasons discussed in Section 1, the assessment aims to provide regionally comparable estimates of maximum NPD potentials for the entire United States for the purpose of policy considerations. The results may be used to identify potential regions and set priorities instead of getting a precise estimate at a specific site. Identification of the most appropriate sites and designs for investment will still rely on developers' efforts.

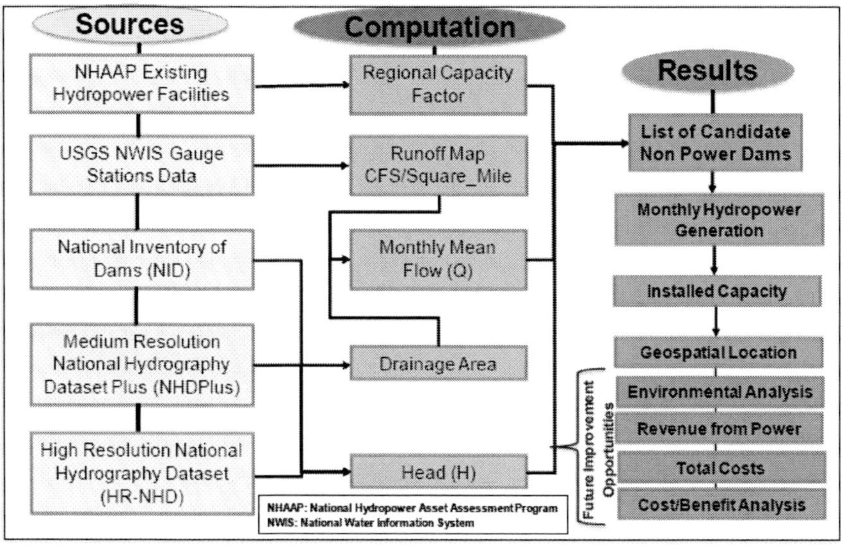

Figure 2. Approach to the NPD resource assessment.

2.2. Data Requirements and Sources for Nationwide Assessment of NPD Potential

As noted above, in order to estimate potential hydropower capacity for NPDs using the electricity generation model, several types of information are needed: monthly flow, head, and monthly capacity factor. Most of the information is not readily available.

A wide variety of data sources were utilized to assemble the information needed for the energy generation model.

2.2.1. USACE NID

NID was utilized in this assessment to provide a comprehensive list of the NPDs and their corresponding dam heights.

The goal of NID is to include all dams in the United States that meet at least one of the following criteria:

1. High hazard classification – loss of one human life is likely if the dam fails.
2. Significant hazard classification – possible loss of human life and likely significant property or environmental destruction if the dam fails.
3. Equals or exceeds 25 feet in height and exceeds 15 acre-feet in storage.
4. Equals or exceeds 6 feet in height and exceeds 50 acre-feet in storage.

Congress first authorized the USACE to inventory dams in the United States with the National Dam Inspection Act (Public Law 92-367) of 1972. The NID was first published in 1975, with a few updates— as resources permitted—over the next 10 years. The Water Resources Development Act of 1986 (P.L. 99-662) authorized USACE to maintain and periodically publish an updated NID, with re-authorization and a dedicated funding source provided under the Water Resources Development Act of 1996 (P.L. 104- 3). USACE also began close collaboration with the Federal Emergency Management Agency (FEMA) and state regulatory offices to obtain more accurate and complete information. The National Dam Safety and Security Act of 2002 (P.L. 107-310) reauthorized the National Dam Safety Program and included the maintenance and update of the NID by USACE. The most recent Dam Safety Act of 2006 reauthorized the maintenance and update of the NID. The latest version of NID, 2007 (http://www.nid.usace.army.mil), contains data about 83,987 dams together with such information as their purpose, location, river name, drainage area, dam height, dam storage, ownership, and primary usage. Although the NID contains most of the dams in the United States, other NPDs that are missing in the database are included in this assessment when identified from the databases described below.

2.2.2. Dams from Reclamation

Data for about 450 dams, along with their characteristics, were collected from Reclamation. Among those dams were more than 100 that were not included in the NID. These dams are merged with NID as a full list of dams for NPD resource assessment.

2.2.3. National Hydropower Asset Assessment Program (NHAAP) Baseline Database

The NHAAP (Hadjerioua et al., 2011) describes the development and construction of the baseline engineering and geospatial information systems and integrated data sets that characterize the hydropower generation inventory in the United States. The NHAAP baseline database is designed to assess and analyze the existing national hydropower infrastructure and provide historical data to study and plan for future potential hydropower upgrades, as well as potential increases in the U.S. hydropower generation.

2.2.4. Natural Resources Conservation Service Watershed Boundary Dataset (WBD)

Watershed boundaries define the aerial extent of surface water drainage to a point. The intent of defining hydrologic units (HU) for the WBD is to establish a drainage boundary framework, accounting for all land and surface areas. The latest version of the WBD defines six HU levels: region (HUC02), subregion (HUC04), basin (HUC06), sub-basin (HUC08), watershed (HUC10), and subwatershed (HUC12). Only regions (HUC02) were used in this report's nationwide assessment.

2.2.5. USGS NHD

The NHD is a comprehensive set of digital spatial data representing the surface water of the United States using common features, such as lakes, ponds, streams, rivers, canals, and oceans. High-resolution (1:24,000-scale) NHDs, covering the entire conterminous United States, were used in the NPD resource assessment (Simley et al., 2009).

2.2.6. NHDPlus Version 1 from USGS and Horizon Systems Corporation

By integrating a variety of datasets—including the National Elevation Dataset (NED), the National Land Cover Dataset (NLCD), and the WBD—into the medium-resolution (1:100,000-scale) NHD, NHDPlus Version 1 adds a variety of useful attributes to NHD features. These attributes include cumulative drainage area characteristics, flow direction, flowline minimum/maximum elevations and slopes, and flow volume and velocity estimates for each flowline in the stream network. (http://www.horizon-systems.com/nhdplus/)

2.2.7. National Water Information System (NWIS) Gauge Observation from USGS

Data were collected from about 22,000 USGS NWIS gauge stations, including location, drainage area, and monthly streamflow observations over the past 30 years. Among those 22,000 stations, 5,595 provide continuous streamflow observations from 1999 to 2008.

2.3. Data Synthesis and Computation of Potential

2.3.1. Initial Dam Selection

The initial selection process identifies dams that are candidates for hydropower development. Existing hydropower impoundments and dams that impound water released through existing powerhouses elsewhere on an impoundment are excluded from this set. Also excluded are dams with erroneous coordinates that could not be resolved through independent investigation of maps and records, as well as dams with erroneous flow or drainage area attributes that could not be corrected though independent investigation. Finally, dams with a reported height of less than five feet were excluded.

2.3.1.1. Update NID Dams Location Using NHD

In order to estimate the NPD flow correctly, a precise dam location is required. However, it was observed that some NID coordinates were not sufficiently accurate and, hence, a correct linkage between NPD and stream segment could not be built. The problem was partially resolved by using high-resolution NHD. High-resolution NHD provides adjusted locations of 41,757 NID dams. In such cases, the dam location is adjusted to be on the high-resolution NHD flowline. After updating NID dam locations, 697 dams were identified as having invalid locations and 267 dams were outside of the conterminous United States, thus they were deleted from the analysis.

2.3.1.2. Exclude Existing Powered Dams, Auxiliary Dams, and Low Head Dams

Based on the NID attributes, existing powered dams, auxiliary dams, and low head dams were excluded.

1. Criteria for selecting existing powered dams from NID: When the "Purpose" field in NID contains "H," which means hydroelectric, the dams are considered existing powered dams. There are 2,569 NID dams identified as existing powered dams. However, it was observed that this NID attribute was not always accurate and up-to-date. The NHAAP database was utilized in the quality control process (Section 2.4) to resolve this issue.
2. Criteria for selecting auxiliary dams from NID: When the "OtherStructureID" field in NID contains some value (i.e., not null), the dams are considered auxiliary dams. Since they share the same headwater with the main dam, it would be an overestimation to include their potential in the full national portfolio. There are 703 NID dams identified as auxiliary dams.
3. Criteria for selecting low head dams from NID: When the "NID_Height" field in NID is missing or less than 5 feet, the dams are considered low head dams. There are 319 NID dams identified as low head dams.

It also should be noted that some dams may fit into multiple criteria. For instance, some auxiliary dams may have an "H" remark in the "Purpose" field. The NPDs that remained in the database after filtering through the three criteria are analyzed further to address these cases.

2.3.1.3. Identification of Dams on Streams with Negligible Streamflow

The spatial analysis process of dam context was performed using NHDPlus (1:100,000 scale) and high-resolution NHD (1:24,000 scale) to ensure accurate registration of dams to stream segments. The high-resolution NHD analysis is essential for correct registration of dams with negligible streamflow (DWNS) that impound very small streams. These streams do not appear in the medium resolution NHDPlus, so exclusive use of NHDPlus to register streams would erroneously assign DWNS to larger streams and overestimate their production potential.

The criteria for identifying DWNS, as long as any one is met, are as follows:

1) A dam is close to a high-resolution NHD flowline, but far from a lower-resolution NHDPlus flowline (Figure 3).
2) A dam is not close to either a high-resolution NHD flowline or an NHDPlus flowline (Figure 4).

There are 24,478 NID dams classified as DWNS. Although there may be some hydropower potential for these types of NPDs, given that there is very limited flow information to support the computation of their potential, they cannot be correctly estimated based on the current data availability. These 24,478 dams could be further analyzed in future small hydropower research efforts to estimate their aggregate potential, though this could be negligible. The remaining 55,707 NPDs can be estimated for their hydropower potential i n the following steps.

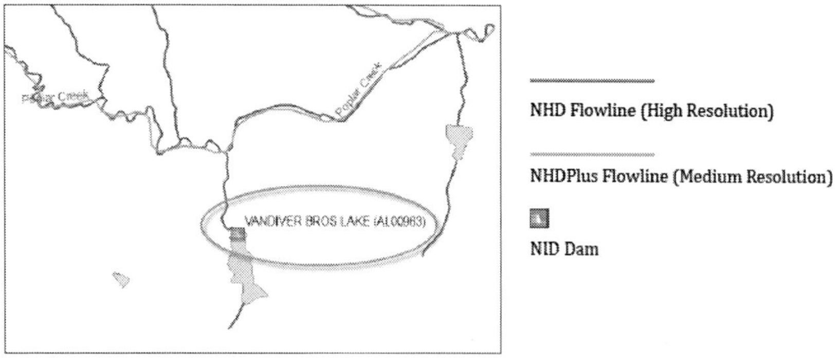

Figure 3. I llustration of the first criterion classification of DWNS – In this example, because the dam "Vandiver Bros Lake" is close to NHD flowline but far from NHDPlus fl owl i ne, it is designated as located on segment with negligible streamflow.

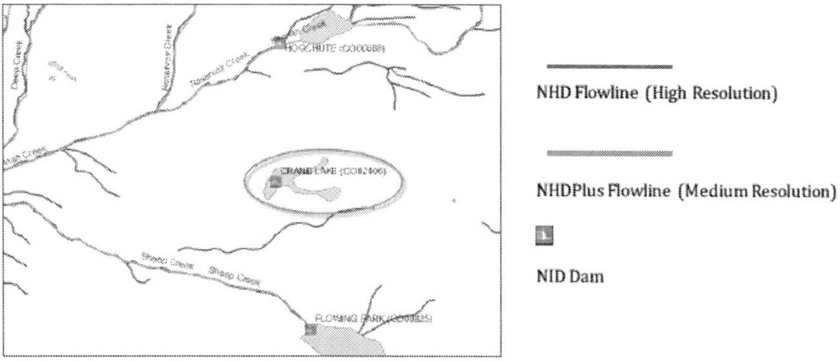

Figure 4. Illustration of the second criterion for selecting DWNS – In this example, because the dam "Crane Lake" is not close to any NHD flowline or NHDPlus flowline, it cannot be linked to any stream and is designated as located on segment with negligible streamfl ow.

2.3.2. Streamflow

Although the most credible approach to estimate available streamflow for hydropower generation is through stream gauge monitoring, many NPDs are not monitored or have streamflow records for a duration of less than 10 years. Some preliminary estimates are available at the national scale, such as the annual mean flow provided in NHDPlus; however, seasonal variability, which is crucial to hydropower production, is usually unavailable. Therefore, alternatives must be sought, especially given the large number of NPDs included in this assessment.

The monthly mean flow was estimated at each of the NPD sites by relying on several available data sources. The key concept of estimating monthly mean streamflow is based on the following equation:

$$\text{Streamflow} = \text{Drainage Area} * \text{Runoff} \tag{4}$$

In Equation (4), "Drainage Area" is the cumulative drainage area (square miles) of the basin above an NPD. "Runoff" is the monthly mean normalized streamflow (cfs/square mile) within an NPD's entire drainage area. Given that the streamflow magnitude is strongly correlated to drainage area, a runoff map has been widely applied in practice to estimate flow at ungauged locations. The monthly runoff at the HUC06 level was estimated in this study to help identify the monthly variability of streamflow. However, this over-simplification may also introduce some errors. A quality control procedure therefore is introduced (described in Section 2.4) to ensure the accuracy of the estimates.

2.3.2.1. Drainage Area

Two sources were utilized to estimate the drainage area for each NPD. The first source is NID. In NID, drainage area is provided for more than half of the dams, but the information sometimes is inaccurate. A better source of drainage area information is NHDPlus. For most of the flowlines (linear features, including stream/river, canal/ditch, pipeline, artificial path, coastline, and connector), NHDPlus provides the cumulative drainage area at the endpoint (outlet) of the flowline. The average length of flowlines in NHDPlus is about 1.2 miles. So, the drainage area of an NPD is approximately identical to the cumulative drainage area of the flowline on which the dam is located. Spatial analyses were performed using NID and NHDPlus stream network to find the flowline on which an NPD is located. Then, the corresponding drainage area provided in NHDPlus was used. This process is described in Figure 5.

2.3.2.2. Runoff

The monthly mean runoff was calculated for all basins (HUC06) based on observation data from 5,595 USGS NWIS gauge stations with continuous streamflow observations between 1999 and 2008. The calculation is an area-weighted average method based on the streamflow observation, location, and drainage area of these 5,595 selected gauge stations. Within each HUC06, the mean runoff for each calendar month is computed from all stations located in that basin.

The annual runoff is shown as an example in Figure 6. As shown, most of the surface water runoff is located in the eastern and northwestern United States.

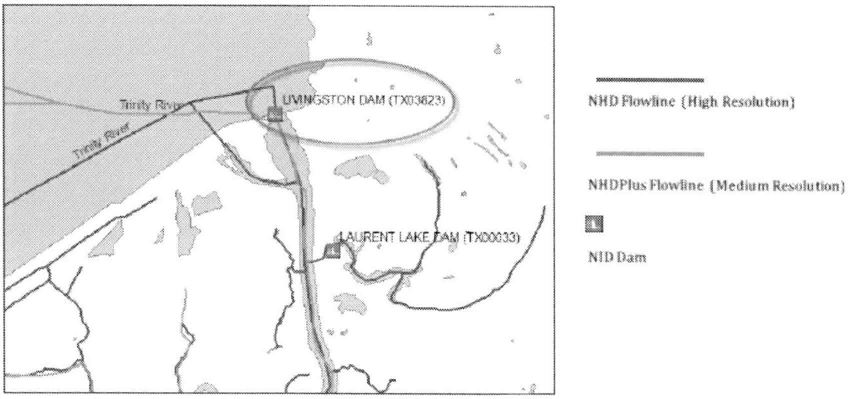

Figure 5. Illustration of linkage between NID and NHDPlus – In NID, Livingston Dam is linked to an NHDPlus flowline (in red color), which represents the Trinity River. The accumulative drainage area for this NHDPlus flowline is 16,572 square miles, which will be used as the dam's drainage area.

2.3.2.3. Streamflow Calculation and Selection

For each NPD, the basin runoff and drainage area are applied in Equation (4) to estimate the monthly streamflow. Because there were two versions of drainage area, alternative estimates of streamflow were computed from the NHDPlus and NID.

The selection of streamflow was based on this criterion: use the streamflow based on drainage area from NHDPlus unless it is at least 50 times greater than the streamflow based on drainage area from NID. The criterion was needed because, owing to inaccuracy of dam locations, the spatial analysis process might mistakenly find the wrong NHDPlus flowline on which that dam is located, which will significantly affect the accuracy of the drainage area retrieved from NHDPlus. As mentioned earlier, an additional quality control step is introduced later to ensure the reasonableness of the flow estimates.

Figure 7 shows the calculated monthly mean streamflow for John Sevier Dam, located in Tennessee, which was derived by this method.

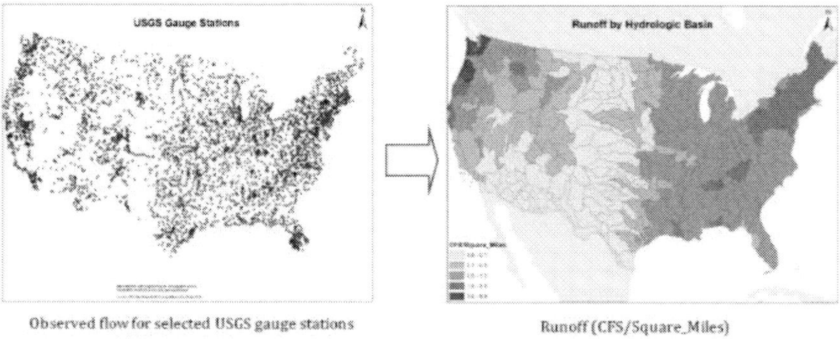

Figure 6. Illustration of runoff derivation.

Figure 7. Monthly mean streamflow time series for John Sevier Dam, Tennessee.

2.3.3. Hydraulic Head

Given the national scope of this study, the choices of penstock and pipeline—which are generally utilized to increase the hydraulic head for hydropower generation—were not considered. The focus is to identify NPD potential with the least new construction and, hence, only the head created by the current existing NPDs was considered. Although the most accurate estimate of hydraulic head is the height difference between headwater and tailwater elevations, such information is not commonly available in the databases. Therefore, the dam height information in NID was utilized as a surrogate to estimate hydraulic head. Four types of height information are provided in NID (graphically depicted in Figure 8), including the following:

1. *Hydraulic Height:* Vertical difference between the maximum design water level and the lowest point in the original streambed.
2. *Dam Height:* Vertical distance between the lowest point on the crest of the dam and the lowest point in the original streambed.
3. *Structural Height*: Vertical distance from the lowest point of the excavated foundation to the top of the dam. Top of dam refers to the parapet wall and not the crest.
4. *NID Height*: Maximum value among Hydraulic Height, Dam Height, and Structural Height.

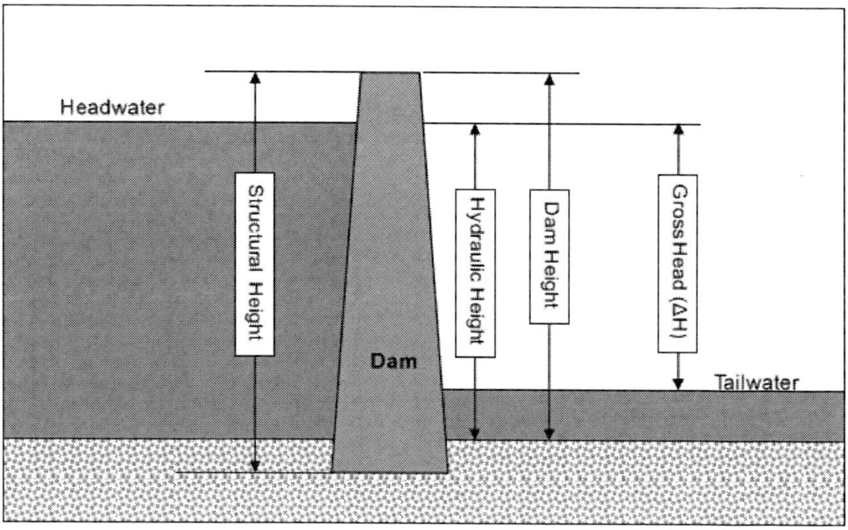

Figure 8. NI D height definitions.

Although it seems that the hydraulic height could be the best choice to represent the gross head, especially for sites with small tailwater depths, this information is not available for many NPDs. However, sometimes the hydraulic height given is equal to the dam height; in these instances, using the hydraulic height will grossly overestimate the computed potential power and, thus, the potential installed capacity.

Therefore, as a "rule of thumb"estimate, 70% of the NID height is used. The following empirical rules estimate the gross head for the NPDs:

1) If Hydraulic Height is not provided, use 0.7 * NID Height.
2) If Hydraulic Height and NID Height are both provided and are equal, use 0.7 * NID Height.
3) If Hydraulic Height is provided but is greater than 0.7 * NID Height, use 0.7 * NID Height.
4) If Hydraulic Height is provided and is less than 0.7 * NID Height, use Hydraulic Height.

In the case of major main-stem NPDs and high dams (over 100 feet), these assumptions were supplemented by analysis of data from navigation lock lift heights and from gauge stations to gain a more realistic estimate of gross head (the difference between headwater and tailwater elevations). For other dams, there could be some high uncertainties between the NID dam height-based estimate and the actual gross head available for power generation. The primary uncertainty in estimating energy potential, based on these NID dam height assumptions, is that the influence of tailwater elevation and the absence of accurate data on tailwater elevation were not included in the current data sources. In addition, the seasonal variability of hydraulic head is not available in most of the cases. The current choices served as preliminary estimates. The 70% (or 0.7 multiplier factor) used with the NID dam height is based on the assumption that about 70% of the NID dam height should be considered as gross head available for power. The tailwater depth and the distance from the upper pool to the dam crest are assumed to be about 30%. This assumption is considered to be reasonable when the NID height is equal to dam height. When the NID height is equal to structural height, since there is an additional excavation depth to be included, the gross estimate head available for power could be overestimated; however, only 2% of the total number of NPDs with 0.4% of the total estimated potential installed capacity was computed using structural dam height in this study. The actual time-varying head at a specific site should be monitored as part of a feasibility study or characterized by detailed pool elevation records from the site operator. Summary for gross head estimation used in this study is shown in Table 2.

2.3.4. Capacity Factor

The regional capacity factor was obtained from the NHAAP database. By using Equation (2), the capacity factor of each existing hydropower plant with available generation and capacity data from 2001 to 2008 was computed (see Figure 9). The plant generation is then utilized as a weighting factor to average the capacity factor for each of the hydrologic regions (HUC02).

Table 2. Type of Heads Used to Estimate Potential Installed Capacity

Estimated Gross Head	Number of NPD	Total Potential Capacity (MW)
Adjusted using Lock and Dam lift from USACE	87	6,921.6
Computed using 0.7 * Dam Height	49859	3,744.2
Adjusted during quality control (NPD > 100 feet)	62	729.9
Using NID Hydraulic Height	3,329	599.8
Using 0.7 * NID Structural Height	1050	44.24
Adjusted from Reclamation Resource Assessment	4	21.5
Total	54,391	12,061.2

Therefore, the regional capacity factor represented the ratio of current generation to the maximum possible generation when generators are constantly operated. The capacity factor was utilized in Equation (3) to estimate the potential NPD capacity.

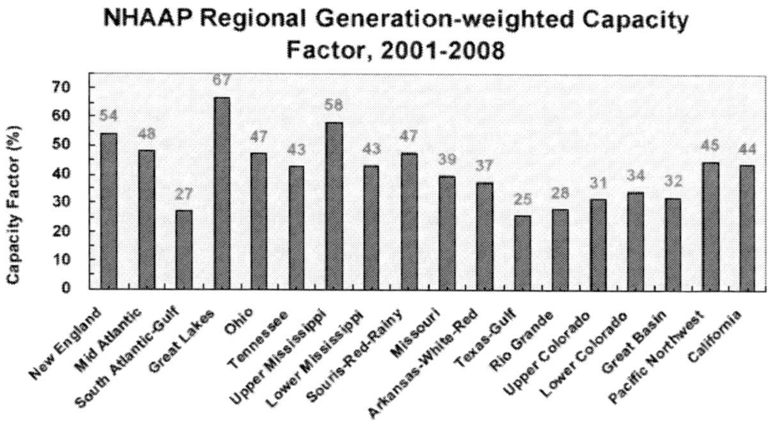

Figure 9. Computed regional capacity factors.

2.3.5. Potential Generation and Capacity

With Equation (1), using hydraulic head and monthly mean streamflow, the mean monthly potential NPD generation was calculated. Mean annual potential NPD generation then can be derived by summing monthly estimates of generation together. The annual potential generation then is fed into Equation (3), along with the regional capacity factor, to estimate the potential capacity. A program was developed to automatically compute the power potential for all NPDs. The preliminary results then were examined through a quality control process.

2.4. Quality Control

Although most of the NID non-powered dams were linked to the NHDPlus stream segments through the algorithm described in Section 2.3, mismatches occasionally occurred. The mismatch typically resulted in unrealistically high flow, which then overestimated the power potential. These errors were usually caused by the following:

- Inaccurate coordinates and attributes of some NID dams
- Different regional hydrologic characteristics
- Inconsistent coordinates between NID and NHD
- Outdated information of some NID attributes
- Overestimated head (mostly for run-of-river dams).

As a result, quality control through manual checking was used to ensure the accuracy of the national estimates. The major quality control steps include the following:

1) *Removal of smaller ponds next to major rivers* – It is a fundamental GIS limitation that the same geographical units may have slightly different coordinates across various datasets. One common solution is to assign a tolerance distance when merging different datasets together (in this case, NID and NHD). However, since many ponds are located near major rivers (e.g., Ohio River), there were many cases that these off-stream ponds were pushed into major rivers and resulted in overestimated power potential. The problem, which was

most common in HUC Region 5 (Ohio), was resolved by examining satellite images.

2) *Removal of existing hydropower plants* – While information was provided in the NID to indicate whether or not a dam was utilized for hydropower generation, such information was not always accurate and up-to-date. The latest hydropower plant inventory was obtained from the NHAAP to correct errors in the NID database. In cases where an NPD shares the same headwater with an existing hydropower dam, the NPD was excluded during the quality control process.
3) *Flow adjustment* – Although two different methods were utilized to estimate the monthly flow in ungauged streams, the most accurate estimates are derived directly from nearby gauge stations. During the quality control, the annual mean flow provided by NHDPlus was utilized as a reference standard to help with validating the flow estimates. If the annual flow estimate was not within a 10% difference of the value in NHDPlus, flow adjustment was considered. In most cases, the monthly flow estimates were rescaled to match the annual mean flow from NHDPlus.
4) *Head adjustment* – Much of the NPD power potential comes from the USACE navigation system (locks and dams). Due to the need of inland river transportation, the flow discharge at locks and dams was always high and can be a potential source of future hydropower. However, the NID dam height may not be reasonably used as a proxy of head in these cases due to the high tailwater elevation. Therefore, the average lift information was obtained from USACE to replace the original head estimate. Although this quality control step resulted in a 7 GW decrease of power potential, the locks and dams remain good candidates, even after the large reduction. In addition to locks and dams, head adjustment also was performed for those high head (> 100 ft) and high potential (> 1 MW) NPD sites. The normal pool head water elevation (usually taken from the dam owner website or the operation documents) and tail water elevation (usually taken from the downstream USGS gauge stations) were identified as the maximum bounding range of the available head. Although this process is considered to be realistic, it is extremely time- and resource-consuming, and elevation measured from different sources may be determined inconsistently. This process resulted in another 0.5 GW reduction of the power potential.

5) *Removal of under-construction sites* – It was noticed that some potentially developable sites already are under construction. For instance, several USACE locks and dams (e.g., Cannelton and Smithland) on the Ohio River are being developed for hydropower generation and, hence, were removed from the candidate list. Several other NPDs—including L&D 52 and 53, Newburgh, and J.T. Myers—are expected to be replaced or converted for hydropower generation in the near future. The national NPD estimate will be adjusted periodically when new information becomes available.
6) Updated from other regional resource assessments – Although flow and head can be estimated through the approach described in the previous section for a large number of NPDs nationwide, these estimates are only preliminary and will not be as accurate as the ones measured on-site. There have been many other regional efforts in evaluating potential hydropower sites in which the designed flow and head were determined based on the actual site conditions. Whenever a credible previous resource assessment becomes available, the information is utilized to improve the national NPD estimates in this study. Part of the Reclamation 2011 information was incorporated to provide more realistic estimate of head.

It should be noted that quality control is a time-consuming process and cannot be performed for all NPDs. Therefore, quality control checks commenced with the NPDs having the highest power estimates; these checks continued gradually following the descending order of power potential. At the current phase of this project, nearly 1,000 NPDs with high hydropower potential have been preliminarily checked.

Not all of the many NPDs analyzed have been subjected to detailed quality control checks for streamflow, head, energy, and capacity estimation. Quality control checking has focused on those dams with the greatest estimated energy production. The raw estimate from Section 2.3 revealed 54,801 NPD sites with potential capacities totaling 43.5 GW. The numbers were reduced to 54,561 NPDs with potential capacities totaling 19.7 GW after steps 1–3 (described in Section 2.4) were applied on part of the NPDs. The most recent estimate included steps 4–5 (described in Section 2.4); with the resulting potential capacity of 12 GW was based on the remaining 54,391 NPDs. While many of these NPDs of lesser potential have not received explicit quality control review, such checks are unlikely to result in substantial increases or decreases in potential capacity because the NPDs with the highest power estimates already have been subjected to the quality control process.

3. RESULTS

3.1. Estimated Energy Availability

Following the methodologies described in Section 2, the potential generation and capacity are estimated for more than 54,000 NPDs in the conterminous United States. These preliminary hydropower potentials consider only approximations of site characteristics (e.g., dam height) and hydrologic variability without further assessment of site feasibility, environmental impact, and economical benefit. At the current phase of assessment, the total potential capacity and annual generation are estimated to be, respectively, 12 GW and 45 terawatt hours (TWh) per year—around 15% of the existing U.S. conventional hydropower total. Current economic limitations on project size will likely produce designs with less capacity than indicated herein.

3.1.1. Nationwide Potential

Although a large number of NPDs are assessed in this study, most of the energy potential is found in a relatively small subset of dams. As illustrated in Figure10, the top 597 NPD sites, each with a potential capacity greater than 1 MW, contribute nearly 90% of the estimated additional national capacity from NPDs. In particular, 40% of the national total is from the top 25 NPD sites only. Therefore, quality control of the top NPD sites was important to ensure the reliability of the national estimate. It should be noted that all sites with potential capacity greater than 1 MW have been examined during the quality control process.

The top 597 NPDs have a potential capacity greater than 1 MW. A regional map of the NPD sites with potential capacity greater than 1 MW is shown in Figure 11. Most of the potential sites are found in the northeastern United States, Ohio River Basin, and Upper and Lower Mississippi River Basins. Specifically, high potentials are found for many USACE locks and dams—87 sites with a total potential of 6.9 GW. The finding is reasonable because the streamflow magnitude must be sufficiently large at locks and dams to support river transportation. Because locks and dams were built mainly for navigation purposes instead of municipal water supply and irrigation, there may be less concern about impacts regarding other competing water usage.

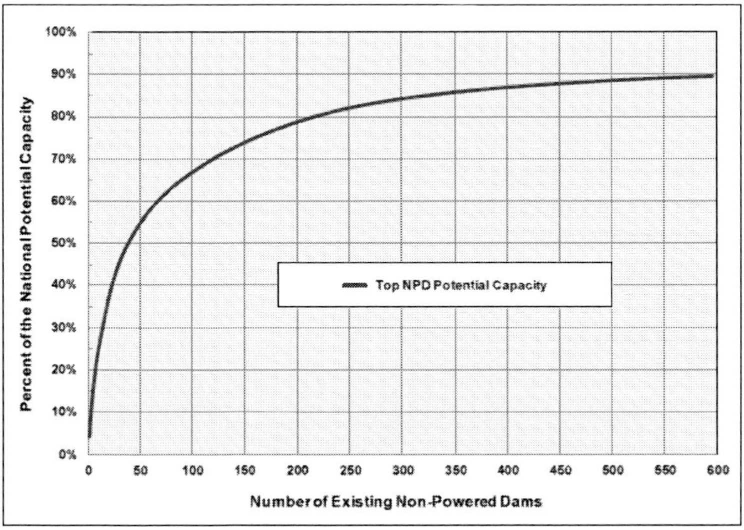

Figure 10. Cumulative potential capacity distribution of the top existing NPDs with the greatest potential added generation capacity.

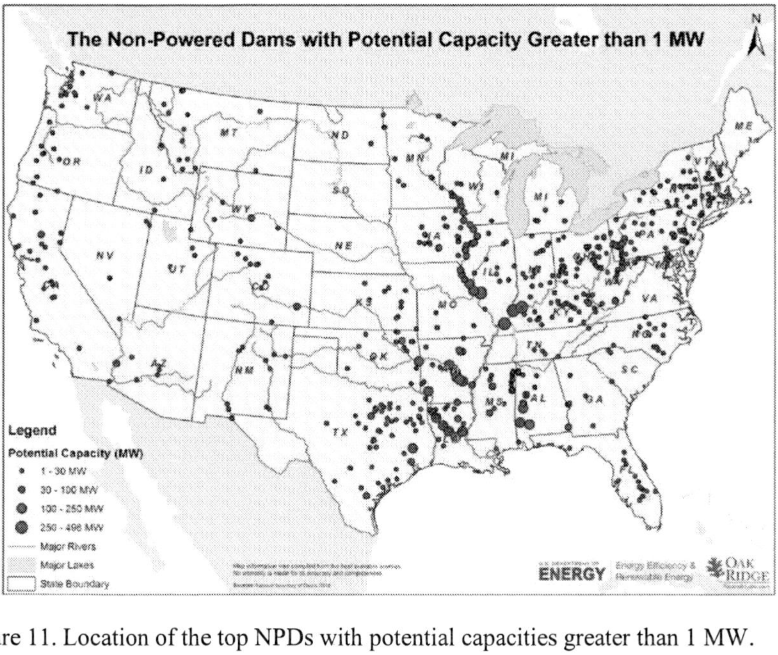

Figure 11. Location of the top NPDs with potential capacities greater than 1 MW.

3.1.2. Regional Potential by River Basin and State

The NPD assessment is summarized in Table 3 by hydrologic regions. The top three regions are highlighted; namely Ohio, Upper Mississippi, and Arkansas-White-Red. As mentioned earlier, most of the hydropower potential identified within these three regions is located at navigation locks and dams located on relatively big rivers. The other summary is shown by state in Table 4. If a NPD was located on the boarder of multiple states, the potential capacity was distributed evenly into each neighboring state to compute the state total. According to this analysis, the greatest amount of hydropower potential is found in Alabama, Arkansas, Illinois, Kentucky, Pennsylvania, Texas, and Louisiana; this mainly is due to a series of Ohio River locks and high river flows. Many of these Ohio River locks are within the top 10 list of the potential NPD sites. It should be noted that other Ohio River locks and dams are not reflected in Table 3 and Table 4, including Cannelton, Smithland, and Meldahl, which were found to be suitable for hydropower development in the initial assessment. They were excluded during the quality control process because construction currently is underway to convert these NPDs for hydropower generation.

Table 3. Summary of NPD Assessment by Hydrologic Regions Totaling 12 GW of Potential

Hydrologic Regions (HUC02)	Potential Capacity (MW)	Potential Generation (TWh/yr)	Hydrologic Regions (HUC02)	Potential Capacity (MW)	Potential Generation (MWh/yr)
1 New England	243	1.110	10 Missouri	258	0.865
2 Mid-Atlantic	479	1.997	11 Arkansas-White-Red	1898	5.960
3 South Atlantic-Gulf	1618	3.778	12 Texas-Gulf	608	1.308
4 Great Lakes	156	0.903	13 Rio Grande	98	0.241
5 Ohio	3236	13.603	14 Upper Colorado	53	0.145
6 Tennessee	53	0.197	15 Lower Colorado	124	0.370
7 Upper Mississippi	2027	9.943	16 Great Basin	29	0.080
8 Lower Mississippi	743	2.802	17 Pacific Northwest	225	0.871
9 Souris-Red-Rainy	58	0.239	18 California	156	0.586

3.2. Integration with Existing Hydropower Inventory and Other Renewable Energy Facilities

Figure 12 illustrates the generation potential listed in Table 3. It is interesting to note that while hydropower is already a significant source of electricity in the Pacific Northwest and California regions, the best potential for new development at NPDs is at locations with less existing hydropower usage; in particular, there was major potential found in the Ohio, Upper and Lower Mississippi, and ArkansasWhite-Red regions. Therefore, new NPD development can help diversify the spatial distribution of national hydropower investment.

More importantly, hydropower is found to be a complementary energy source with other renewables.

Table 4. Summary of NPD Assessment by State Totaling 12 GW of Potential

State	Potential Capacity (MW)	State	Potential Capacity (MW)	State	Potential Capacity (MW)
AL	922	ME	19	OH	288
AZ	80	MD	48	OK	339
AR	1136	MA	67	OR	116
CA	195	MI	48	PA	679
CO	172	MN	186	RI	13
CT	68	MS	271	SC	38
DE	3	MO	489	SD	12
FL	173	MT	88	TN	40
GA	144	NE	7	TX	658
ID	12	NV	16	UT	40
IL	1269	NH	63	VT	17
IN	454	NJ	33	VA	50
IA	427	NM	103	WA	85
KS	92	NY	295	WV	210
KY	1253	NC	167	WI	245
LA	857	ND	31	WY	45

A comparison is shown in Figure 13. By overlaying the potential NPD sites with National Renewable Energy Laboratory (NREL) estimates of wind and solar power potentials (NREL, 2011), it can be seen how the development of hydropower at NPDs may complement the development of other renewable technologies and augment the national portfolio. While more wind and solar power potentials are found in the western and southwestern United States, hydropower development through NPDs can provide clean renewables for other regions in the nation.

3.3. Future Improvement Opportunities

During the execution of the NPD assessment, the study team identified several refinements that were not possible within the current scope of the effort but that may yield improved estimates.

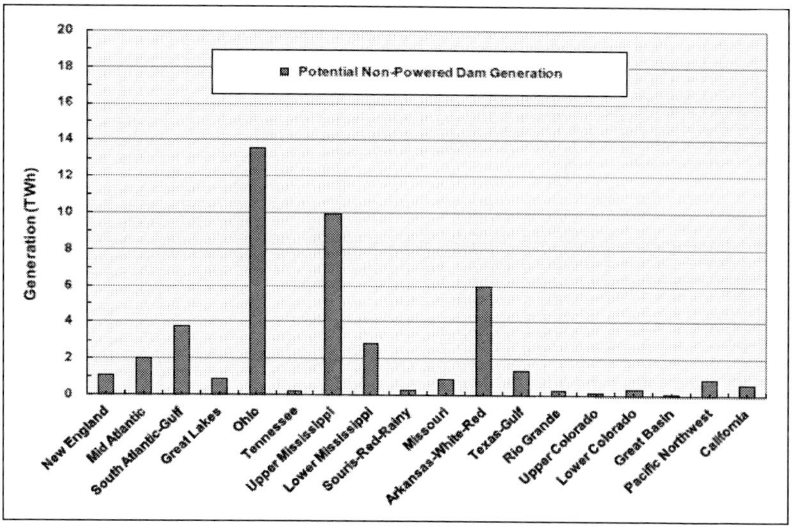

Figure 12. Potential hydropower generation from existing NPDs.

Foremost is the application of refined site development models that include economic constraints on turbine and power house capacity based on the frequency of occurrence of daily flows. Such refinement will yield estimates of energy production that are more realistic, but less than the total energy available at a site. In the context of refined site development modeling, there are dams in high mountain regions where an alternative development scheme with a diversion penstock may yield more hydraulic head—and greater energy production—than the simple dam height or pool elevation difference modeling provides.

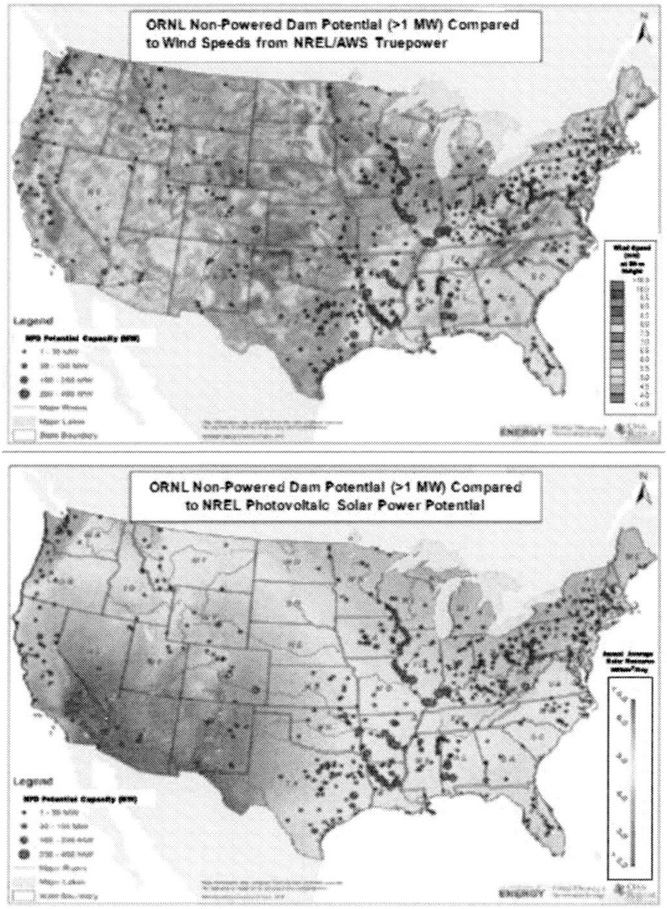

Figure 13. Comparison of top NPD sites with potential capacities greater than 1MW with maps of wind and solar photovoltaic resource potential.

Enhanced modeling of water availability through collaboration with USGS research efforts will yield more robust estimates that consider long-term climate variability and provide more accurate seasonal statistics and flow-duration relationships for use in energy analysis. Future efforts could examine environmental, socioeconomic, and electric power infrastructure attributes of the NPD population to estimate site feasibility and development cost in addition to potential.

Appendix A. Top 100 NPD with Hydropower Potential

Estimated Annual Average

	Dam Name	***Owner name*	*City*	*County*	*State*	*River Name*	*Year comp-leted*	*Estimated Head (feet)*	*Flow (cfs)*	**Generation (MWh)*	*Estimated Potential *Capacity (MW)*
1	OHIO RIVER LOCKS & DAM 53	CELRL	OLMSTED	PULASKI	KY / IL	OHIO	1929	12.0	276,990	2,084,907	496.0
2	OHIO RIVER LOCKS & DAM 52	CELRL	BROOKPORT	MASSAC	KY / IL	OHIO	1929	12.0	276,133	2,078,454	494.5
3	JOHN T. MYERS LOCKS & DAM	CELRL	UNIONTOWN	UNION	KY / IN	OHIO	1975	18.0	147,134	1,661,216	395.2
4	NEWBURGH LOCKS & DAM	CELRL	SCUFFLETOWN	HENDERSON	KY / IN	OHIO	1975	16.0	133,554	1,340,348	318.9
5	MELVIN PRICE LOCKS & DAM	CEMVS	ALTON	MADISON	IL / MO	MISSISSIPPI	1990	24.0	97,508	1,467,886	299.3
6	COFFEEVILLE LOCK & DAM	CESAM	COFFEEVILLE	CLARK	AL	TOMBIGBEE	1962	34.0	26,470	564,511	241.7
7	DEMOPOLIS LOCK & DAM	CESAM	DEMOPOLIS	MARENGO	AL	TOMBIGBEE	1955	40.0	22,113	554,822	237.6
8	CLAIBORNE LOCK & DAM	CESAM	CLAIBORNE LANDING	MONROE	AL	ALABAMA	1969	30.0	28,340	533,297	228.3
9	RED RIVER W.W. LOCK & DAM 03	CEMVK	COLFAX	NATCHITOCHES	LA	RED	1991	31.0	30,192	587,080	187.0
10	DAVID D. TERRY LOCK & DAM	CESWL	PINE BLUFF	PULASKI	AR	ARKANSAS	1968	18.0	45,857	517,745	164.9
11	JOE HARDIN LOCK & DAM	CESWL	GRADY	LINCOLN	AR	ARKANSAS	1968	20.0	48,420	607,436	161.0
12	COL CHARLES D. MAYNARD LOCK & DAM	CESWL	REDFIELD	JEFFERSON	AR	ARKANSAS	1968	17.0	45,970	490,195	156.1
13	LIVINGSTON DAM	TRINITY RIVER AUTHORITY	LIBERTY	SAN JACINTO	TX	TRINITY RIVER	1969	70.0	7,453	327,233	152.0
14	LOCK & DAM 25	CEMVS	WINFIELD	LINCOLN	MO / IL	MISSISSIPPI	1939	15.0	76,764	722,259	147.3
15	LOCK & DAM 24	CEMVS	CLARKSVILLE	PIKE	MO / IL	MISSISSIPPI	1940	15.0	76,366	718,512	146.5

	Dam Name	**Owner name	City	County	State	River Name	Year comp-leted	Estimated Head (feet)	Flow (cfs)	*Generation (MWh)	Estimated Potential *Capacity (MW)
16	TOAD SUCK FERRY LOCK & DAM	CESWL	CONWAY	FAULKNER	AR	ARKANSAS	1969	16.0	45,336	454,988	144.9
17	RUSSELL B. LONG LOCK & DAM	CEMVK	GRAND ECORE	NATCHITOCHES	LA	RED	1995	25.0	28,663	449,465	143.1
18	W.D.MAYO LOCK & DAM	CESWT	FORT COFFEE	LE FLORE	OK	ARKANSAS	1970	21.0	32,145	423,426	134.9
19	EMMETT SANDERS LOCK & DAM	CESWL	PINE BLUFF	JEFFERSON	AR	ARKANSAS	1968	14.0	46,007	404,007	128.7
20	JOE D. WAGGONNER, JR. LOCK & DAM	CEMVK	SIMMESPORT	CATAHOULA	LA	RED	1985	25.0	25,242	395,825	126.1
21	JOHN OVERTON LOCK & DAM	CEMVK	SIMMESPORT	RAPIDES	LA	RED	1987	24.0	31,322	471,522	125.0
22	JONESVILLE LOCK & DAM	CEMVK	SIMMESPORT	CATAHOULA	LA	BLACK	1972	30.0	24,577	462,472	122.6
23	MILLWOOD DAM	CESWL	ASHDOWN	LITTLE RIVER	AR	LITTLE	1966	74.0	6,818	316,458	100.8
24	MONTGOMERY LOCKS & DAM	CELRP	MONACA	BEAVER	PA	OHIO	1936	18.0	37,166	419,626	99.8
25	MISSISSIPPI RIVER DAM 22	CEMVR	SAVERTON	RALLS	MO / IL	MISSISSIPPI	1938	10.0	74,224	465,570	94.9
26	MISSISSIPPI RIVER DAM 21	CEMVR	QUINCY	ADAMS	MO / IL	MISSISSIPPI	1938	10.0	72,781	456,518	93.1
27	MONTGOMERY POINT LOCK & DAM	CESWL	WATSON	DESHA	AR	WHITE	2004	20.0	27,982	351,041	93.0
28	MISSISSIPPI RIVER DAM 20	CEMVR	CANTON	LEWIS	MO / IL	MISSISSIPPI	1936	10.0	72,126	452,410	92.2
29	LOCK & DAM 15	CEMVR	ROCK ISLAND	ROCK ISLAND	IL / IA	MISSISSIPPI	1934	16.0	42,841	429,950	87.7
30	EMSWORTH LOCKS & DAMS	CELRP	PITTSBURGH	ALLEGHENY	PA	OHIO	1938	18.0	31,424	354,796	84.4
31	JOHN MARTIN DAM & RESERVOIR	CESPA	LAS ANIMAS	BENT	CO	ARKANSAS	1943	78.5	5,027	247,622	78.9
32	LOCK & DAM 18	CEMVR	GLADSTONE	HENDERSON	IL / IA	MISSISSIPPI	1937	10.0	54,831	343,930	70.1
33	DUBUQUE NUMBER 11	DAEN NCR (CORPS OF ENGR)	DUBUQUE	DUBUQUE	IA	MISSISSIPPI	1937	12.0	44,801	337,215	68.8
34	RED ROCK DAM	CEMVR	KNOXVILLE	MARION	IA	DES MOINES	1969	104.0	4,995	325,849	66.4
35	MISSISSIPPI RIVER DAM 13	CEMVR	CLINTON	WHITESIDE	IA / IL	MISSISSIPPI	1939	11.0	46,987	324,201	66.1

Appendix A. (Continued)

	Dam Name	***Owner name*	*City*	*County*	*State*	*River Name*	*Year completed*	*Estimated Head (feet)*	*Flow (cfs)*	**Generation (MWh)*	*Estimated Potential *Capacity (MW)*
36	OOLOGAH LAKE	CESWT	CLAREMORE	ROGERS	OK	VERDIGRIS	1963	86.0	3,502	188,933	60.2
37	MISSISSIPPI RIVER DAM 14	CEMVR	LECLAIRE	SCOTT	IA / IL	MISSISSIPPI	1939	11.0	42,773	295,121	60.2
38	MISSISSIPPI RIVER DAM 16	CEMVR	MUSCATINE	ROCK ISLAND	IA / IL	MISSISSIPPI	1937	9.0	48,050	271,253	55.3
39	GEORGE W ANDREWS LOCK & DAM	CESAM	GORDON	HOUSTON	AL / GA	CHATTAHO O CHEE	1963	25.0	8,203	128,638	55.1
40	PALO VERDE DIVERSION	DOI BR	MAYFLOWER	YUMA,	AZ / CA	COLORADO	1957	46.0	5,603	161,670	54.3
41	WILLIAM BACON OLIVER REPLACEMENT	CESAM	TUSCALOOSA	TUSCALOOSA	AL	BLACK WARRIER	1992	28.0	7,159	125,742	53.8
42	MISSISSIPPI RIVER DAM 12	CEMVR	BELLEVUE	JACKSON	IA / IL	MISSISSIPPI	1938	9.0	45,165	254,971	52.0
43	COLUMBIA LOCK & DAM	CEMVK	RIVERTON	CALDWELL	LA	OUACHITA	1970	18.0	17,301	195,334	51.8
44	A.I.SELDEN	CESAM	EASTPORT	HALE	AL	BLACK WARRIOR	1958	22.0	8,625	119,016	51.0
45	MISSISSIPPI RIVER DAM 17	CEMVR	NEW BOSTON	MERCER	IA / IL	MISSISSIPPI	1939	8.0	48,144	241,588	49.3
46	JOHN C. STENNIS	CESAM	COLUMBUS	LOWNDES	MS	TOMBIGBEE	1978	27.0	6,588	111,567	47.8
47	DASHIELDS LOCKS & DAM	CELRP	CORAPOLIS	ALLEGHENY	PA	OHIO	1929	10.0	31,559	197,956	47.1
48	TYGART DAM	CELRP	GRAFTON	TAYLOR	WV	TYGART	1938	133.5	2,360	197,513	47.0
49	LOCK & DAM 08	CEMVP	GENOA	VERNON	WI / MN	MISSISSIPPI	1937	11.0	32,842	226,599	46.2
50	BLUESTONE DAM	CELRH	HINTON	SUMMERS	WV	NEW	1947	48.0	6,255	188,328	44.8
51	LOCK & DAM 10	CEMVP	GUTTENBERG	CLAYTON	IA	MISSISSIPPI	1937	8.0	43,261	217,083	44.3
52	ROSS BARNETT RESERVOIR DAM	PEARL RIV. VAL. WTR. SUP. DIST	JACKSON	RANKIN	MS	PEARL	1965	44.8	3,678	103,343	44.2
53	ALLEGHENY LOCK & DAM 03	CELRP	HARMAR TOWNSHIP	ALLEGHENY	PA	ALLEGHEN Y	1934	14.0	20,954	184,003	43.8
54	WESLEY E SEALE DAM	CITY OF CORPUS CHRISTI	CORPUS CHRISTI	JIM WELLS	TX	NUECES	1958	62.3	2,323	90,775	42.2

	Dam Name	**Owner name	City	County	State	River Name	Year comp-leted	Estimated Head (feet)	Flow (cfs)	*Generation (MWh)	Estimated Potential *Capacity (MW)
55	FELSENTHAL LOCK & DAM	CEMVK	STERLINGTON	UNION	AR	OUACHITA	1978	18.0	13,488	152,284	40.4
56	LOCK & DAM 09	CEMVP	HARPERS FERRY	ALLAMAKEE	IA	MISSISSIPPI	1937	9.0	33,817	190,908	38.9
57	DE CORDOVA BEND DAM	BRAZOS RIVER AUTHORITY	PECAN PLANTATION	HOOD	TX	BRAZOS	1969	79.0	1,626	80,558	37.4
58	FISH BARRIER	CDWR	OROVILLE	BUTTE	CA	FEATHER	1964	61.0	3,444	131,765	35.2
59	ALLEGHENY LOCK & DAM 02	CELRP	SHARPSBURG	ALLEGHENY	PA	ALLEGHEN Y	1934	11.0	21,133	145,812	34.7
60	ALLEGHENY LOCK & DAM 04	CELRP	NATRONA	ALLEGHENY	PA	ALLEGHEN Y	1927	11.0	20,741	143,111	34.0
61	LOCK & DAM NO 05	DAEN NCS (CORPS OF ENGR)	MINONA	BUFFALO	WI / MN	MISSISSIPPI	1935	9.0	29,155	164,588	33.6
62	LOCK & DAM NO 01	CITY OF BATESVILLE	BATESVILLE	INDEPENDENCE	AR	WHITE	1904	14.0	11,329	99,482	31.7
63	DEWEY DAM	CELRH	AUXIER	FLOYD	KY	JOHNS CREEK OF LEVISA FORK	1949	86.0	2,469	133,165	31.7
64	LOCK & DAM 07	CEMVP	LA CRESCENT	WINONA	MN / WI	MISSISSIPPI	1937	8.0	30,942	155,269	31.7
65	ALLEGHENY LOCK & DAM 07	CELRP	KITTANNING	ARMSTRONG	PA	ALLEGHEN Y	1931	13.0	16,313	133,017	31.6
66	LOCK & DAM NO 02	ARKANSAS POWER AND LIGHT COMPANY	TICHNOR	ARKANSAS	AR	WHITE	1905	14.0	11,070	97,207	31.0

Appendix A. (Continued)

	Dam Name	***Owner name*	*City*	*County*	*State*	*River Name*	*Year comp-leted*	*Estimated Head (feet)*	*Flow (cfs)*	**Generation (MWh)*	*Estimated Potential *Capacity (MW)*
67	LAKE HOUSTON DAM	CITY OF HOUSTON	MAGNOLIA GARDENS	HARRIS	TX	SAN JACINTO	1954	62.0	1,698	66,021	30.7
68	LOCK & DAM NO 03	ARKANSAS COLLEGE	GRADY	INDEPENDENCE	AR	WHITE	1908	14.0	10,960	96,247	30.7
69	GRAY REEF	USBR	ALCOVA	NATRONA	WY	NORTH PLATE	1961	30.0	5,390	101,423	30.2
70	GRENADA DAM	CEMVK	GRENADA	GRENADA	MS	YALOBUSH A	1954	97.0	1,868	113,654	30.1
71	FRANKLIN FALLS DAM	CENAE	FRANKLIN	MERRIMACK	NH	PEMIGEWA S SET	1943	112.0	1,961	137,757	30.1
72	MAXWELL LOCKS & DAM	CELRP	EAST MILLSBORO	FAYETTE	PA	MONONGA H ELA	1964	20.0	9,815	123,133	29.3
73	ENFIELD DAM	C.H. DEXTER	ENFIELD	HARTFORD	CT	CONNECTIC UT	1825	12.6	16,940	133,884	29.3
74	CHATFIELD DAM	CENWO	DENVER	DOUGLAS	CO	SOUTH PLATTE	1973	124.0	1,257	97,750	29.1
75	RODMAN DAM AND SPILLWAY	FL-DEP	WELAKA	PUTNAM	FL	OKLAWAHA	1968	38.0	2,710	64,598	27.7
76	BELTON LAKE	CESWF	BELTON	BELL	TX	LEON	1954	117.3	804	59,153	27.5
77	HOWARD A HANSON DAM	CENWS	KANGLEY	KING	WA	GREEN	1962	149.5	1,084	101,620	26.3
78	CHARLEROI LOCKS & DAM	CELRP	MONESSEN	WESTMORELAN D	PA	MONONGA HELA	1967	17.0	10,309	109,926	26.2
79	COON RAPIDS	THREE RIVERS PARK DISTRICT	COON RAPIDS	HENNEPIN	MN	MISSISSIPPI	1913	23.0	8,773	126,569	25.8
80	LOCK & DAM 04	CEMVP	ALMA	BUFFALO	WI / MN	MISSISSIPPI	1935	7.0	28,207	123,849	25.3
81	MORELOS DIVERSION	IBWC	YUMA	YUMA	AZ / CA	COLORADO	1950	18.0	6,641	74,985	25.2

	Dam Name	***Owner name*	*City*	*County*	*State*	*River Name*	*Year comp-leted*	*Estimated Head (feet)*	*Flow (cfs)*	**Generation (MWh)*	*Estimated Potential *Capacity (MW)*
82	NEW SAVANNAH BLUFF LOCK & DAM	CESAS	AUGUSTA	RICHMOND	GA	SAVANNAH	1937	15.0	6,127	57,645	24.7
83	SAYLORVILLE DAM	CEMVR	DES MOINES	POLK	IA	DES MOINES	1975	94.2	2,021	119,400	24.3
84	B. EVERETT JORDAN DAM	CESAW	HAYWOOD	CHATHAM	NC	HAW	1974	61.0	1,460	55,846	23.9
85	LOCK C-1 DAM AT WATERFORD	NYS CANAL CORP	WATERFORD	SARATOGA	NY	HUDSON	1912	16.8	9,359	98,625	23.7
86	ADAM T. BOWER MEMORIAL	DCNR	SHAMOKIN DAM	SNYDER	PA	SUSQUEHA NNA	1969	5.6	27,692	97,271	23.3
87	LOCK & DAM 06	CEMVP	TREMPEALEAU	TREMPEALEAU	WI / MN	MISSISSIPPI	1936	6.0	29,754	111,980	22.8
88	BARREN RIVER LAKE DAM	CELRL	BOWLING GREEN	BARREN	KY	BARREN	1964	118.0	1,267	93,766	22.3
89	LA GRANGE LOCK & DAM	CEMVR	LA GRANGE	CASS	IL	ILLINOIS	1939	10.0	17,304	108,540	22.1
90	ABERDEEN LOCK & DAM	CESAM	ABERDEEN	MONROE	MS	TOMBIGBEE	1981	27.0	3,037	51,436	22.0
91	EAGLE AND PHENIX MILL LAKE DAM	FIELDCREST MILLS, INC.	COLUMBUS	MUSCOGEE	GA / IL	CHATTAHO OCH EE	1851	9.8	8,309	51,074	21.9
92	LOCK 32 DAM ERIE CANAL	NYS CANAL CORP	PITTSFORD	MONROE	NY	NEW YORK STATE BARGE CANAL	1908	28.0	7,039	123,631	21.3
93	KENTUCKY RIVER LOCK & DAM 05	COMMONWE ALTH OF KENTUCKY	LAWRENCEBE RG	ANDERSON	KY	KENTUCKY	1844	22.0	6,384	88,093	21.0
94	WISTER LAKE	CESWT	WISTER	LE FLORE	OK	POTEAU	1949	95.0	1,099	65,513	20.9
95	STEVE BOONE	STEVE BOONE	NO TOWN	HENRY	KY	TRIB-KENTUCKY	1992	18.2	7,569	86,412	20.6
96	GOLD RAY DAM	JACKSON COUNTY PARKS & RECREATION	GOLD HILL	JACKSON	OR	ROGUE	1941	35.0	3,589	78,785	20.4

Appendix A. (Continued)

	Dam Name	***Owner name*	*City*	*County*	*State*	*River Name*	*Year comp-leted*	*Estimated Head (feet)*	*Flow (cfs)*	**Generation (MWh)*	*Estimated Potential *Capacity (MW)*
97	SUTTON DAM	CELRH	SUTTON	BRAXTON	WV	ELK	1960	109.9	1,238	85,289	20.3
98	GREEN RIVER LOCK & DAM 02	CELRL	CALHOUN	MCLEAN	KY	GREEN	1956	14.0	9,690	85,091	20.2
99	KENTUCKY RIVER LOCK & DAM 08	COMMONWE ALTH OF KENTUCKY	BRYANTSVILL E	GARRARD	KY	KENTUCKY	1900	25.0	5,392	84,553	20.1
100	ARKABUTLA DAM	CEMVK	COLDWATER	CALLOWAY	MS	COLDWATE R	1943	84.0	1,433	75,493	20.0

*Estimated Generation and Capacity are based on the total energy available at the site.

**Legend for Owner Name column is located in Appendix B.

APPENDIX B. ACRONYMS

CDWR	California Department of Water Resources
CENAB	Corps of Engineers North Atlantic Baltimore District
CENAN	Corps of Engineers North Atlantic New York District
CENAO	Corps of Engineers North Atlantic Norfolk District
CENAP	Corps of Engineers North Atlantic Philadelphia District
CENAE	Corps of Engineers North Atlantic New England District
CENAU	Corps of Engineers North Atlantic Europe District
CENWK	Corps of Engineers Northwestern Kansas District
CENWO	Corps of Engineers Northwestern Omaha District
CENWP	Corps of Engineers Northwestern Portland District
CENWS	Corps of Engineers Northwestern Seattle District
CENWW	Corps of Engineers Northwestern Walla Walla District
CEPOA	Corps of Engineers Pacific Ocean Alaska District
CEPOF	Corps of Engineers Pacific Ocean Far East District
CEPOH	Corps of Engineers Pacific Ocean Honolulu District
CEPOJ	Corps of Engineers Pacific Ocean Japan District
CESAC	Corps of Engineers South Atlantic Charleston District
CESAJ	Corps of Engineers South Atlantic Jacksonville District
CESAM	Corps of Engineers South Atlantic Mobile District
CESAS	Corps of Engineers South Atlantic Savannah District
CESAW	Corps of Engineers South Atlantic Wilmington District
CESPA	Corps of Engineers South Pacific Albuquerque District
CESPL	Corps of Engineers South Pacific Los Angeles District
CESPK	Corps of Engineers South Pacific Sacramento District
CESPN	Corps of Engineers South Pacific San Francisco District
CESWF	Corps of Engineers Southwestern Forth Worth District
CESWG	Corps of Engineers Southwestern Galveston District
CESWL	Corps of Engineers Southwestern Little Rock District
CESWT	Corps of Engineers Southwestern Tulsa District
CELRB	Corps of Engineers Great Lakes and Ohio River Buffalo District
CELRC	Corps of Engineers Great Lakes and Ohio River Chicago District
CELRE	Corps of Engineers Great Lakes and Ohio River Detroit District
CELRH	Corps of Engineers Great Lakes and Ohio River Huntington District
CELRL	Corps of Engineers Great Lakes and Ohio River Louisville District

CELRN	Corps of Engineers Great Lakes and Ohio River Nashville District
CELRP	Corps of Engineers Great Lakes and Ohio River Pittsburgh District
CEGRC	Corps of Engineers Gulf Region Central District
CEGRN	Corps of Engineers Gulf Region North District
CEGRS	Corps of Engineers Gulf Region South District
CEMVM	Corps of Engineers Mississippi Valley Memphis District
CENVN	Corps of Engineers Mississippi Valley New Orleans District
CEMVR	Corps of Engineers Mississippi Valley Rock Island District
CEMVS	Corps of Engineers Mississippi Valley St. Louis District
CEMVP	Corps of Engineers Mississippi Valley St. Paul District
CEMVK	Corps of Engineers Mississippi Valley Vicksburg District
DOE	Department of Energy
DOI	U.S. Department of the Interior
DOI BR	U.S. Department of the Interior Bureau of Reclamation
DWNS	Dams With Negligible Streamflow
ECRE	Eagle Creek Renewable Energy
EIA	Energy Information Administration
FERC	Federal Energy Regulatory Commission
HU	Hydrologic Unit
NED	National Elevation Dataset
NHA	National Hydropower Association
NHAAP	National Hydropower Asset Assessment Project
NHD	National Hydrography Dataset
NID	National Inventory of Dams
NLCD	National Land Cover Dataset
NPD	National Non-Powered Dam
NREL	National Renewable Energy Laboratory
NWIS	National Water Information System
ORNL	Oak Ridge National Laboratory
P	Precipitation
Q	Runoff
TVA	Tennessee Valley Authority
USACE	U.S. Army Corps of Engineers
USBR	U.S. Bureau of Reclamation (or, Reclamation)
USGS	U.S. Geological Survey
WBD	Watershed Boundary Dataset

REFERENCES

U.S. Department of the Interior, et al. 2007. Potential Hydroelectric Development at Existing Federal Facilities, for Section 1834 of the Energy Policy Act of 2005.

Energy Information Administration (EIA). 2010. From EIA-906, EIA-920, and EIA-923 Databases. http://eia.doe.gov/cneaf/ electricity/page/eia906_920.html

Hadjerioua, B., S.-C. Kao, M. J. Sale, Y. Wei, S. K. SanthanaVannan, H. A. Shanafield III, D. P. Kaiser, R. Devarakonda, C. Odeh, G. Palanisamy and B. T. Smith. 2011. National Hydropower Asset Assessment Project. Oak Ridge National Laboratory, U.S. Department of Energy.

Hall, D. G., S. J. Cherry, K. S. Reeves, R. D. Lee, G. R. Carroll, G. L. Sommers, and K. L. Verdin. 2004. Water Energy Resources of the United States with Emphasis on Low Head/Low Power Resources, DOE/ID-11111, April 2004.

National Renewable Energy Laboratory. 2011. U.S. Solar Resource Maps (Photovoltaic). Obtained from http://www.nrel.gov/gis/solar.html

U.S. Bureau of Reclamation. 2011. Hydropower Resource Assessment at Existing Reclamation Facilities. Denver, CO. March 2011.

Simley, J. D., W. J. Carswell Jr. 2009. The National Map—Hydrography: U.S. Geological Survey Fact Sheet 2009-3054, 4p.

End Note

[1] 1 gigawatt (GW)=1,000 megawatts (MW). On an annual basis, 1 MW of hydropower produces enough electricity to power nearly 400 U.S. homes. Each gigawatt could power up to 400,000 homes.

INDEX

D

E

F

G

S

T

U

V

W